RAND NATIONAL SECURITY RESEARCH DIVISION

Security Sector Reform in Ukraine

Olga Oliker, Lynn E. Davis, Keith Crane, Andrew Radin,
Celeste Ward Gventer, Susanne Sondergaard, James T. Quinlivan,
Stephan B. Seabrook, Jacopo Bellasio, Bryan Frederick,
Andriy Bega, Jakub Hlavka

Prepared for the Ukraine Investment Alliance

For more information on this publication, visit www.rand.org/t/RR1475-1

Library of Congress Cataloging-in-Publication Data is available for this publication.
ISBN: 978-0-8330-9597-8

Published by the RAND Corporation, Santa Monica, Calif.

Cover: Ukrainian soldiers march on Independence Square in downtown Kiev (photo by Danil Shamkin/NurPhoto via AP Images).

The RAND Corporation is a research organization that develops solutions to public policy challenges to help make communities throughout the world safer and more secure, healthier and more prosperous. RAND is nonprofit, nonpartisan, and committed to the public interest.

RAND's publications do not necessarily reflect the opinions of its research clients and sponsors.

Preface

This report provides a series of recommendations for the reform of Ukraine's security and defense institutions. This research was undertaken in response to a request by the presidential administration of Ukraine and in participation with the National Security and Defense Council and sponsored by Ukraine Investment Alliance, a 501(c)(4) foundation. Research for this report was completed in the fall of 2015. Although some minor updates have been made, the analysis predominantly reflects the situation as of that time.

This report should be of interest to those in Ukraine who are engaged in security sector reform and those in the international community supporting such reform in Ukraine. This report can also be useful to those interested in assisting with security sector reform in other countries.

The research was conducted within the International Security and Defense Policy Center of the RAND National Security Research Division (NSRD). NSRD conducts research and analysis on defense and national security topics for the U.S. and allied defense, foreign policy, homeland security, and intelligence communities and foundations and other nongovernmental organizations that support defense and national security analysis.

For more information on the International Security and Defense Policy Center, see www.rand.org/nsrd/ndri/centers/isdp or contact the director (contact information is provided on the web page).

Contents

Figures and Tables

Figures

Tables

Summary

The Maidan Revolution in Ukraine created an opportunity for change and reforms in a system that had resisted them for a quarter of a century. The war in the eastern region of Donbass that began in 2014 and continues to this day has highlighted the desperate need for reforms both within Ukraine's armed forces and in the security sector more broadly.

In this report, we look at several aspects of Ukraine's security sector. We assess what different institutions need to do and where gaps exist that preclude these institutions from being effective, efficient, transparent, and accountable. We then provide recommendations for changes Ukraine could make that would improve existing practices, in line with Euro-Atlantic standards and approaches. We provide recommendations that do not require constitutional changes, due to the political challenges inherent in such reforms. However, we note that constitutional changes may prove advisable to ensure that Ukraine can build the security architecture that will serve it best.

Clarify Roles and Responsibilities

The roles and responsibilities of the President and Cabinet of Ministers (CoM) are ambiguous and the CoM is unwieldy. There are gaps and overlaps in the functions performed by the Ministry of Defense (MoD) and the General Staff (GS). Civilian control remains weak below the President and CoM. The Chief of Defense Force (CHoD) reports directly to the President, thereby disempowering the Minister of Defense.[1] Regulations and organizational culture tend to push routine decisions upward to at least the Deputy Minister level, contributing to a culture of avoiding responsibility.

In the absence of substantial constitutional reform, which could prove politically infeasible, we recommend clarifying the roles and responsibilities of Ukraine's leadership through legislation and policy documents. In particular, we recommend these declared roles and responsibilities:

President: The President as the commander-in-chief would assume responsibility for the security and defense of Ukraine against threats to its sovereignty and independence. This description of the President's role reflects the Constitution's requirement that the President "administer the national security and defence of the State."[2] The President would have responsibility for the command and control of military operations and policy control over the MoD and, through the Minister of Defense, over the Armed Forces of Ukraine.

Prime Minister/Cabinet of Ministers: The Prime Minister and CoM would direct and coordinate all ministries, take the lead in the budget process, and ensure that Ukraine's laws are implemented. Through its functions as the government of Ukraine, the CoM would have operational and policy control of all ministries except Defense, and direction of financial, personnel, and related activities for all ministries including Defense.

Ministry of Defense: The MoD would be responsible for administering the Armed Forces of Ukraine, as well as for military command and control, including command of all forces deployed for military contingencies in the territory of Ukraine. The CHoD, GS, and Armed Forces of Ukraine would be subordinate to the Minister. All military operations would be subordinate to a Joint Operational Command (JOC). Forces from other agencies deployed to any current military operations led by the Armed Forces of Ukraine would be under the exclusive command and control of the commander of the relevant

---

[1] We use Chief of the Defense Force as the title for the head of the General Staff. The position in Ukraine today is called Chief of the General Staff.

[2] Constitution of Ukraine, Article 106, Kyiv: Supreme Council of Ukraine, 1996.

operation under the JOC, although administratively, forces from other agencies remain part of their agency.

Improve Coordination Across the Security Sector

In Ukraine, ministries and agencies operate independently, with little accountability and coordination. They have their own resources, make their own decisions, and set their own tasks. Sharing of information is inconsistent. Individual ministries or agencies fail to specialize and instead invest in broad ranges of capabilities. Decisionmaking is often taken to the highest levels, overwhelming senior officials. Organizations designed to coordinate ministries and agencies are weak or ineffective. The National Security and Defense Council (NSDC) lacks budgetary authority and the authority to ensure implementation of its policy decisions.

To better coordinate the activities of the key ministries and agencies, we recommend expanding the authorities and capabilities of the NSDC to ensure that it can provide not only a forum for coordination but also the ability to ensure that the President's decisions are executed on a day-to-day basis. Ukraine needs to change informal practices and create a culture of greater cooperation across departments and agencies. Senior leaders need to make clear their desires that information be shared, departments and agencies collaborate, and decisionmaking be delegated to the lowest level possible.

The presidential administration should elevate the status of the Joint Committee on Intelligence (JCOI) under the NSDC and establish the position of Chairman of the JCOI as the principal intelligence adviser to the President, CoM, and NSDC. The JCOI would serve as the overall head of the Ukrainian intelligence community, responsible for coordinating its roles, missions, budgets, and activities.

We recommend the creation of a new interagency cyber coordination committee, the Joint Committee on Cyber Security under the auspices of the NSDC, modeled on our recommendation for the JCOI. The Joint Committee on Cyber Security would be responsible for coordinating Ukraine's cybersecurity activities, including the development and continuous review of a national cybersecurity strategy and concept.

A critical problem in the eyes of providers of assistance to Ukraine is the lack of coordination. We recommend in the short term the creation of an ad hoc "Board to Coordinate Foreign Defense Assistance" as part of the NSDC.

Undertake Changes in Structures and Organizations

We recommend a number of structural reforms of tasks and functions within the existing ministries and agencies with the goals of clarifying roles and responsibilities, reducing duplication, improving account-

Table S.1
Recommended Structural Reforms

Security Sector	Recommended Change
Internal Security	• Redefine the tasks of the Security Service of Ukraine as a domestic intelligence organization, with more clearly and narrowly defined authorities, to increase accountability and coordination with other intelligence agencies. • Continue the reform plan of the Ministry of Internal Affairs (MIA) to make it a small supervisory organization responsible for directing and coordinating a range of separate state agencies. • Organize the National Guard as a European-style gendarmerie under the MIA to improve its flexibility and efficiency. • Maintain State Border Guard Service as a nonmilitarized law enforcement body under the MIA, supported by other organizations as necessary.
Ministry of Defense	• Empower the Minister of Defense as senior civilian adviser to the President, Prime Minister, and Ukraine's parliament (Verkhovna Rada) on defense policies, with a single chain of command wherein the CHoD reports to the President through the Minister. • Create a Deputy Defense Minister, Secretary General, and six functional departments: Personnel, Defense Intelligence Policy, Capability Development, Strategy and Policy, Procurement, and Finance/Comptroller. • Establish the CHoD as the primary military adviser to the Minister and the President, with responsibility for the conduct of military operations and for manning, training, and equipping the Armed Forces of Ukraine. • Reorganize the GS and subordinate units into traditional J-code functions (J-1-8) to align Ukraine with North Atlantic Treaty Organization (NATO) standards.

Table S.1—Continued

Security Sector	Recommended Change
Cybersecurity	• Move the national Computer Emergency Response Team out from under the State Services for Special Communication and Information Protection so that it is a fully autonomous organization responsible for all civilian aspects of cyber incident management and response. • Create a cyber command with capability to conduct full-spectrum cyberspace operations with responsibility for achieving the overall coordination for cyber defense within the MoD and the Armed Forces of Ukraine.
Defense Technical Cooperation	• Facilitate defense imports outside of UkrOboronProm by granting explicit authority for foreign procurement to the MoD. • Improve transparency, efficiency, and competitiveness of UkrOboronProm by making UkrOboronProm's holdings in its subsidiaries and affiliates a matter of public record; incorporating unconsolidated subsidiaries and all auxiliary companies as public companies; and, after careful review, deciding whether to privatize, liquidate, or retain state-controlled enterprises in the defense industry. • Improve strategic trade controls by increasing the government's legal authority over trade in dual-use and defense technologies.

ability, and increasing coordination within and among organizations (Table S.1).

Defense Reforms to Improve Warfighting and Efficient Use of Resources

We provide recommendations in four critical areas that have particular importance for Ukraine's warfighting and efficient use of resources (Table S.2): command, control, communications, and intelligence, where unity of command and secure communications in the current military operations in the Anti-Terror Operation have been lacking; recruiting and training personnel, where the current salary and benefits structure does not efficiently attract and retain personnel; procurement, where contracting is sole-source and affected by conflicts of interest and lack of quality control; and logistics, where significant

Table S.2
Recommendations to Improve Warfighting

Critical Area	Recommendations
Command, control, communications, and intelligence	• Pass legislation to clarify the operational chain of command and create a JOC to centralize responsibilities for ongoing military operations. • Specify roles and responsibilities to devolve authority and ensure that orders do not skip echelons. • Give full command authority over other state organizations participating in military operations to the Armed Forces of Ukraine. • Procure more-secure communications networks by purchasing new equipment.
Recruiting and training personnel	• Reduce the number of senior officers by creating a rank structure similar to Western militaries. • Phase out mobilization and conscription. • Simplify bonuses and compensation for contract soldiers.
Procurement	• Reduce sole-source contracts by reviewing the classification of defense orders. • Competitively bid contracts. • Adopt NATO standards for equipment and supplies.
Logistics	• Adopt a computerized inventory management system. • Set broader and more-flexible supply and equipment requirements (norms).

limitations exist in supplying combat units and parallel supply chains operate among the organizations fighting in the Anti-Terror Operation.

Conclusions

The Ukrainian security organizations that existed in March 2014 were unable to respond effectively to the emerging conflict in Eastern Ukraine. Since March 2014, the Ukrainian security establishment has made significant progress, including improving logistics and pursuing reform of the MIA. Nevertheless, these efforts have been insufficient to address the current and future threats facing Ukraine. Ukraine's security sector needs substantial additional reforms to enable it to become effective, efficient, transparent, and accountable.

This report defines a road map for security sector reform by providing a range of recommendations for organizational change in line with Euro-Atlantic standards and approaches. We believe these recommendations offer a significant improvement over what currently exists, and will be robust across a range of contingencies.

Implementing these reforms will be extremely challenging. The international community can provide continued assistance, but implementation lies in the hands of the Ukrainian government. In the end, the effectiveness of the reforms depends not only on putting appropriate institutions in place, but also on affecting sustainable cultural shifts, which may well prove even more challenging and will require strong leadership throughout Ukraine's national security establishment and its government as a whole.

CHAPTER ONE

Introduction

The Maidan Revolution in Ukraine created an opportunity for changes and reforms in a system that had resisted them for a quarter of a century. The war in the eastern region of Donbass that began in 2014 and continues to this day has highlighted the desperate need for reforms both within Ukraine's armed forces and in the security sector more broadly.

For Ukraine to succeed, not just in that conflict but as an effective and democratic state, it needs a security sector that provides for the security of the Ukrainian people, one that does not waste their tax dollars, one that can be trusted, and one that is effective. This applies to the police officer on the street, the soldier on the front lines, intelligence personnel, and all those in the relevant ministries and agencies that oversee their work, purchase their equipment, and decide where they will go and when. This is something Ukraine has not historically had, and it is something Ukraine unquestionably must build.

There exists no single model for how to build a security sector that is efficient, effective, transparent, and accountable. There are as many models for this as there are states that do it, and each of them changes and adapts as requirements shift. However, given Ukraine's stated goal to adhere to Euro-Atlantic standards, it is plausible to look to the states of the North Atlantic Treaty Organization (NATO) Alliance and the European Union (EU) for approaches to security sector reform. Even here, there are a range of options and no one-size-fits-all solution.

In this report, we look at several aspects of Ukraine's security sector. We assess what different institutions need to do and where gaps exist that preclude these institutions from being effective, efficient, transparent, and accountable. We then provide recommendations for changes Ukraine could make that would improve existing practices, in line with Euro-Atlantic standards and approaches. This analysis is not comprehensive. We have not considered all aspects of the security sector and our recommendations do not touch on every component of even the ministries and agencies we study. It is meant to identify some of the most important problems that need attention in the short term, and that can set the stage for continuing reforms and progress.

Our recommendations will be most relevant for Ukrainians who are striving to define their nation's path forward as it continues to make critical choices for its future. However, both our approach and some of the central recommendations should also be useful to others looking to foster reform—including those in other states who seek to make their security sectors better and their advisers from abroad who are looking for ways to adapt Western (or other) models to the needs of countries in transition.

Approach

The President of Ukraine and the National Security and Defense Council (NSDC) asked the RAND Corporation to develop and provide recommendations for the reform of Ukraine's security sector.[1] The organizations included in Ukraine's security sector in accordance with this remit include the police and other internal security forces, the Armed Forces of Ukraine, intelligence organizations, security services, border control, and organizations responsible for cybersecurity.[2]

---

[1] The RAND study focused on five tasks: overall national security sector architecture, the Ministry of Defense (MoD), intelligence coordination, cybersecurity, and defense-technical cooperation with global partners.

[2] There are other security-related organizations (such as the prosecutors' office and prisons) that are part of the security sector and deserving of study, but they were beyond the scope of this project. On the scope of Ukraine's security sector, see O. Reznikova and V. Tsiukalo, *Development of Ukrainian National Security Strategic Planning and Forecasting System*, Kyiv, Ukraine: National Institute for Strategic Studies, June 2015.

This report is our response. We began our project by identifying three strategic goals for Ukraine's security sector, and the organizations within it. First, security organizations need to provide effective protection against Ukraine's current and future threats, both internal and external. Second, these organizations need to make efficient use of resources, including by reducing corruption. Third, security organizations should be aligned with key Euro-Atlantic standards, including democracy, respect for human rights, rule of law, civilian control of the military, and accountability. We then defined key areas for reform, focusing on where problems exist in meeting these three overarching goals. The first area involved the overall security sector architecture and included the roles and responsibilities of the President and Prime Minister, ways to improve coordination, and the structure of the internal security organizations. The second area focused on reform of the MoD and the General Staff (GS). The third area involved defense reforms to improve warfighting and the efficient use of resources. The fourth area involved cybersecurity and the fifth area focused on defense technical cooperation with global partners.

For each of these areas, our analysis followed a similar logic. First, we outlined what the system should do in a given area, drawing in part on how similar systems function in other countries.[3] Second, we developed an assessment of how well various functions were being performed in Ukraine, based on reviews of pertinent Ukrainian documents, existing literature, and discussions with knowledgeable Ukrainian officials, experts, military personnel, and civil servants, and with foreign advisers to Ukraine. As part of our research, we made many trips to Ukraine, including visits to forces stationed in Eastern Ukraine.

We developed recommendations to correct the problems we observed. Many of these recommendations were based on observations

---

[3] The comparison countries included well-functioning institutions with ties to the Euro-Atlantic alliance, such as Canada, France, Germany, the United States, the United Kingdom, and Australia; countries that faced the challenges of moving from Soviet or Warsaw Pact institutions to NATO-compatible institutions, such as the Czech Republic and Poland; countries with somewhat comparable presidential-parliamentary political systems, such as France and Poland; and countries facing significant security threats, such as Israel, South Korea, and Poland.

of effective practices in other countries. While informed by Ukraine's current organization and its needs, the recommendations for the MoD and cybersecurity were developed as an ideal (yet practicable) structure. The recommendations do not require changes in the Ukrainian Constitution, but implementation of many of them would require radical changes and new legislation. Some of our recommendations identify problems that can be addressed quickly, while others point to more-complex reforms that will require further analysis and discussion. What our report provides is a road map for the reform of Ukraine's security sector.

CHAPTER TWO

Reforming the Security Sector—Overall Architecture

Reform of the security sector in Ukraine had been a challenge since its independence in 1991, and many of the same factors that undermined reform in the past remain a concern even after the Maidan Revolution. While there have been major improvements on the Soviet-era institutions that Ukraine initially inherited, Ukraine's current institutions still reflect a hybrid between those and approaches that reflect Western governance models. Among other factors, Ukraine's path to Western-style political institutions has been impeded by shifts in the perspectives of the governing coalition and by pervasive corruption and clientalism. For example, after the 2004 Orange Revolution, Ukraine's constitution was amended that same year to empower the Prime Minister and Cabinet of Ministers (CoM) and reduce the risk of the President assuming authoritarian control. After Viktor Yanukovitch was elected in 2010, the constitution was again amended to give the President greater authority and avoid debilitating conflicts between the Prime Minister and the President. A key demand of the Maidan Revolution was to return to the 2004 constitution, but this has not yet taken place. Today, there remains continuing uncertainty about the roles of the President and Prime Minister.[1]

Reform has also been hindered by pervasive corruption and clientalism. With rapid privatization, wealth was concentrated into the hands of relatively few individuals, known as "oligarchs," who play an outsized role in politics and governance, pursuing activities such

[1] Mikhail Minakov, "A Decisive Turn? Risks for Ukrainian Democracy After the Euromaidan," Carnegie Endowment for International Peace, Washington, D.C., February 3, 2016.

as funding political movements and gaining control of government institutions to enhance their own wealth. The influence of oligarchs and weakness of Ukraine's institutions facilitated the development of "unwritten rules" that operate in parallel with the written code and ensure that state institutions often work for private interests.[2] As a result, Ukraine's government institutions are weak, and a culture of obstruction, legalism, and secrecy has consistently undermined public trust and the creation of Western-style institutions.[3]

Today the prospects for reform are uncertain. President Petro Poroshenko's government has committed itself to reform, and many activists who had participated in the Maidan protests also remain politically active and insistent on the need for reform. But Ukraine faces many challenges, including continuing entrenched interests vested in the current system.

Although some may argue that another challenge to reform is the continued conflict in Eastern Ukraine, we counter that, in fact, this conflict makes reform more imperative. Other countries, the United States among them, have instituted substantial security sector reforms while at war. Sometimes, such reforms are the only means to victory, and failure to implement them can result in defeat.

In this chapter, we focus on the overall security sector architecture and how it should be designed to respond to

1. the ambiguities in the constitution with respect to the roles and responsibilities of the President and Prime Minister that undermine executive level decisionmaking
2. the lack of coordination among agencies and ministries in the development of security policies
3. the highly decentralized intelligence system
4. the continuing deficiencies in the internal security organizations.

---

[2] Steven Levitsky and Lucan Way, *Competitive Authoritarianism: Hybrid Regimes After the Cold War*, New York: Cambridge University Press, 2010, pp. 218–220; Robert W. Orttung, "What Hinders Reform in Ukraine?" George Washington University Elliott School of International Affairs, Washington, D.C., PONARS Eurasia Policy Memo No. 166, September 2011; Minakov, 2016.

[3] Orttung, 2011.

Clarify Roles and Responsibilities

What Does the System Need to Do?

For effective executive-level decisionmaking, the roles and responsibilities of senior officials need to be clear, especially in a system that shares power between a president and prime minister. Countries differ in how they define the respective roles and these often change over time.

Problems in Ukraine's Current System

In Ukraine, ambiguities in the constitution undermine executive-level decisionmaking in the security sector. According to the constitution, the President is charged with "administering" Ukraine's national security. At the same time, the CoM, led by the Prime Minister, is charged with "directing and coordinating" the country's national security. The result has been competition between the President and the Prime Minister for a decade, and more recently, challenges to reform and effective management. The current system also contributes to stovepiping of key ministries. The leaders are typically aligned with the President or Prime Minister and sometimes reluctant to cooperate with other organizations.

The situation is further exacerbated by legislation. On the one hand,

> the President of Ukraine exercises control of the Armed Forces and other State Military Organisations responsible for national security, defence and law enforcement through powers vested in him/her as the Chairperson of the National Security and Defence Council of Ukraine and if necessary through supporting institutions established in accordance with Article 106, Section One, paragraph 28 of the Constitution of Ukraine.[4]

On the other hand,

> The Cabinet of Ministers, through its constitutional powers, implements State domestic and foreign policies that ensure sovereignty, defence capability, national security, public order and the

[4] Law of Ukraine, "On Democratic Civilian Control of State Military Organisation and Law Enforcement Bodies," Article 13.2, 2003.

fight against criminal activity in accordance with the Constitution, Laws and Presidential Decrees.[5]

Recommendations

Absent constitutional reform, which is beyond the scope of our analysis, we recommend clarifying the roles and responsibilities within Ukraine's existing governance structure through legislation and policy documents. The proper allocation of responsibilities begins at the highest levels of government, but clear roles and responsibilities are needed within the ministries and agencies as well. Formal rules to support decisionmaking by lower levels are also needed to address the problem that virtually all decisions are pushed to higher levels. In particular, we recommend declared roles and responsibilities for the President and for the Prime Minister and CoM.

The President, as the commander-in-chief, would be responsible for the security and defense of Ukraine against threats to its sovereignty and independence. This description of the President's role reflects the Constitution's requirement that the President "administer the national security and defence of the State."[6] It highlights the President's role in responding to major threats to the country while leaving responsibility for day-to-day internal security matters with the CoM and the Ministry of Internal Affairs (MIA). The President would have responsibility for the command and control of military operations, along with policy control over the MoD and, through the Minister of Defense, the Armed Forces of Ukraine.

The Prime Minister and CoM would direct and coordinate all ministries, take the lead in the budget process, and ensure that Ukraine's laws are implemented. Through its functions as the government of Ukraine, the CoM would have operational and policy control of all ministries except Defense, and direction of financial, personnel, and related activities for all ministries including Defense. The CoM, in coordination with the NSDC (of which the key Ministers

---

[5] Law of Ukraine, "On Democratic Civilian Control of State Military Organisation and Law Enforcement Bodies," Article 15, 2003.

[6] Constitution of Ukraine, Article 106, 1996.

are members), would have the responsibility to allocate funds to match policy priorities. Improved budgetary transparency and prioritization would be particularly critical for Ukraine to be able to use its limited resources more efficiently. Improving the capacity of the Ukrainian parliament (Verkhovna Rada) to oversee financial and administrative policy, supported by civil society, would provide an important motivation for greater coordination and efficiency, and could build greater trust in the security services throughout Ukrainian society.

Strengthen Coordination Among Ministries and Agencies

What Does the System Need to Do?

Coordination across different ministries and agencies is a challenge for all governments and is typically achieved through a combination of different mechanisms. Effective systems tend to be ones in which the executive (in this case, the President or CoM) selects and holds accountable the leadership of the security sector agencies. Within this framework, a national security council or similar structure can offer a forum for discussion, consultation, and coordination and can help to ensure that policies are implemented. Formal rules can clearly define the roles and responsibilities of different organizations to reduce uncertainty, inefficiency, and competition. Systems will also be more effective when they take into account existing informal practices and the bureaucratic culture of states and organizations. There may also be situations where changes in bureaucratic culture will be necessary to ensure organizations' ability to coordinate their activities, effectively distribute resources, and plan for the future.

Problems in Ukraine's Current System

Currently, ministries and agencies operate independently, with little accountability and coordination. They have a tendency to act as separate fiefdoms, with their own resources, decisionmaking procedures, and tasks. Although the major security sector organizations are formally accountable to the senior leadership, which is able to enforce some level of coordination in executing key tasks, individual agencies do not appear to take the initiative to perform tasks together. Sharing information or resources, which is critical to effective collaboration, is inconsistent. Diffusion of and uncertainty about responsibility and accountability between the President, the Prime Minister, and the CoM also undermine coordination.

Current coordination problems result in inefficiencies, waste, and loss of trust. Individual ministries or agencies fail to specialize; instead, they invest in broad ranges of capabilities. The National Guard and State Border Guard Service possess separate logistics chains, independent of the Ukrainian army's, to supply their forces near the front, and are increasingly procuring their own armored vehicles and heavy weapons. The MoD has its own strategic-level intelligence capability, in part because the Security Service of Ukraine (SBU) does not always share intelligence. The SBU is able to focus on its own priorities and does not need to coordinate with other organizations—partly because it has extensive resources, including its own tactical units. It is also not well integrated with Ukraine's other intelligence agencies and organizations. As a result, there is often friction when one organization is called on to support the operations of another.

Organizations designed to coordinate ministries and agencies are weak or ineffective. While some formal systems specify how coordination should take place and even exist to facilitate that coordination, they appear to be insufficient. The mandate of the NSDC is to "coordinate and control the activities of executive power bodies in the area of national security and defence,"[7] but it lacks budgetary authority and the capability to ensure the implementation of policy decisions. In theory, a vote of the NSDC compels the President to issue a decree to enforce action. Even if the NSDC could invoke the President's authority, ministries and agencies must also answer to the CoM.

The CoM, representing the Ukrainian government, is the main venue for interagency coordination. As such, it is unwieldy and its ability to act as an efficient coordinating or decisionmaking body is limited. Indeed, it is legally required to approve any decisions involving more than one agency, such as Ukraine's defense equipment orders

---

[7] Constitution of Ukraine, Article 107, 1996.

(and, at times, decisions as minor as what to do about tainted food purchases). This requirement means that the CoM has to vote on an enormous number of decisions, rather than having administrative questions addressed at lower levels. In some ways, this system is one manifestation of a problem that is widespread throughout Ukraine's government: the culture of pushing decisions to the top.

Recommendations

The NSDC needs to be given more responsibility to provide a forum in which the ministries and agencies can voice their views, and to ensure on a day-to-day basis that the President's decisions are implemented.

Ukraine also needs to change informal practices and create a culture of greater cooperation across departments and agencies. Senior leaders should make clear that information should be shared and departments and agencies should collaborate. Steps should be taken to share staff among ministries and agencies through regular rotations, and through the creation of liaison positions and ad hoc coordinating committees in specific policy areas. Decisionmaking should be delegated to the lowest level possible. Civilians from outside the government should be hired and promoted, especially individuals with Western training and experience.

Improve Intelligence Coordination

What Does the System Need to Do?

Coordinating intelligence community activities and sharing intelligence products requires significant and continuous management to be effective. Responsibility for policy oversight and guidance in the intelligence process ultimately resides at the very top, with the President. But a responsible office needs to be designated to ensure effective management, integration, and coordination of intelligence community activities. The degree to which this office and the official who heads it are empowered and have the necessary tools to facilitate coordination will determine the effectiveness of both the office and its director.

Problems in Ukraine's Current System

The intelligence community in Ukraine is highly decentralized and composed of the Foreign Intelligence Service of Ukraine, the Defense Intelligence Agency, and the Intelligence Service of the Ukrainian State Border Guard Service. The SBU and elements within other agencies also conduct intelligence activities, but they are part of the informal intelligence sector.

Oversight and coordination of the Ukrainian intelligence community has evolved over the course of Ukraine's post–Cold War history. A Joint Committee on Intelligence (JCOI) at the NSDC was first established in the early 1990s and evolved over time, although it was later disbanded. In March of 2015, President Poroshenko reestablished the JCOI. Thus far, it has not been able to achieve integration and coordination of the intelligence agencies. A set of policies and procedures is not in place to help coordinate the various intelligence activities, set requirements, or establish links between policymakers and intelligence analysts. The JCOI also lacks sufficient staff.

Recommendations

The presidential administration should elevate the status of the JCOI, clarifying its role through decree or executive order. The decree should establish the position of Chairman of the JCOI as principal intelligence adviser to the President, the CoM, and the NSDC. The JCOI would be authorized and tasked to improve *coordination* of intelligence operations, collection, and analysis across all intelligence organizations (regardless of their current reporting lines) and to increase *sharing* of intelligence information and products across organizational boundaries.

The Committee would serve as the overall head of the Ukrainian intelligence community, responsible for coordinating its roles, missions, budgets, and activities. It would also ensure that the needs of combat forces in the current operation are included in intelligence requirements, and that appropriate mechanisms are in place for disseminating the intelligence. The Committee should set such standards for the intelligence community as training requirements for personnel, pay and benefits scales that are equitable across the enterprise, and information technology protocols.

The staff of the JCOI needs to be expanded, drawing primarily on personnel from the intelligence agencies, but also from outside (e.g., from the volunteer movement). To avoid creating a new bureaucracy, the JCOI staff should be compact and composed of high-quality professionals chosen from the various intelligence organizations. The JCOI should establish "national intelligence managers" among the staff, organized along the lines of the priority needs of users, with responsibility for removing obstacles to coordination and sharing across the intelligence community in meeting those needs.

Align Roles and Functions of Internal Security Organizations

What Does the System Need to Do?

Internal security organizations in Europe and the United States are based on the rule of law and democratic principles. This is accomplished while having the capabilities to effectively respond to domestic threats. Among other critical practices, domestic intelligence organizations are kept small, allocated limited resources, and given specified authorities as a way of limiting the risks of abuse and human rights violations. Achieving coordination and information-sharing are more important than the specific characteristics of an organizational structure.

Problems in Ukraine's Current System

Existing internal security organizations include the MIA, police, SBU, State Border Guard Service, State Bodyguard, Emergency Services, Migration Service, National Guard, State Special Transport Service, and the State Services for Special Communication and Information Protection (SSSCIP). These organizations have undergone substantial reform and redesign since the Maidan Revolution. Many of the internal security agencies were misused by the former regime, and were seen as responsible for repression rather than for providing security and defending democracy. Among the reforms was the creation of the

National Guard out of the former Internal Troops of the MIA and the reorganization of police within the MIA. However, even with these reforms, internal security organizations in Ukraine continue to be insufficient to Ukraine's needs.

The role of the MIA in the supervision and oversight of the security sector remains unclear. The control of organizations within and subordinate to the MIA has changed greatly. Until early 2014, the MoD directed and coordinated the State Emergency Service. Responsibility for the State Emergency Service was then given to the MIA. The MIA has put forward a plan under which it would become a small supervisory organization responsible for directing and coordinating a range of separate state agencies (including the State Border Guard Service, the State Migration Service, the State Emergency Service, and the National Guard). However, the implementation of this reform plan is incomplete and its future uncertain.

The SBU is not in line with European models in terms of its mandate, operations, or size. The SBU developed from the Soviet Union's Committee for State Security—or, as it is more commonly known, the KGB—and has gone through significant reforms. The Ukrainian government has separated border control and foreign intelligence from the SBU. However, in Ukraine's recent history, the leadership has periodically used the SBU to monitor political foes and shore up power, and it continues to report directly to the President, perhaps a reminder of its role in protecting the administration rather than ensuring internal security. The absence of a similar organization in the Euro-Atlantic region reinforces the need to consider reform of the SBU. Indeed, the closest model of the SBU has been the Federal Security Service of the Russian Federation.

The SBU's legal mandate is the protection of the "state security" of Ukraine.[8] It has authority to act both through its mandate to conduct "counterintelligence activity," which is broadly defined to include the "prevention, timely identification and repulsion of external and

[8] Law of Ukraine, "On the Security Service of Ukraine," Article 1, 1992.

internal threats" to the Ukrainian state,[9] and through its law enforcement mandate to investigate certain serious crimes. While many other European domestic intelligence and security agencies have broad mandates to protect their countries against foreign threats, the SBU's use of a vague "counterintelligence" mandate is incompatible with Euro-Atlantic standards. By including "external and internal threats" in the SBU's counterintelligence function (even when these threats are not responses to the intelligence services of its adversaries), the law masks the SBU's role and undermines democratic accountability. There is also a risk that such threats could again become synonymous in the eyes of the government with defense of a regime, with clear negative consequences for Ukraine's democracy. Finally, intelligence functions that the SBU provides are not integrated into an interagency coordination process because the SBU is not considered a formal part of Ukraine's intelligence community.

To pursue its mission, the SBU is legally mandated to have 27,000 personnel and can increase to 31,000 in "special periods," such as circumstances of martial law.[10] Although the purpose and function of the SBU do not conform to those of other European organizations, it often compares itself to the U.S. Federal Bureau of Investigation. However, the SBU is proportionately much larger than the Federal Bureau of Investigation, which has approximately 35,000 personnel for a U.S. population seven times larger than Ukraine's. Similarly, the United Kingdom's MI-5 has approximately 4,000 personnel, although the United Kingdom has 50 percent more people than Ukraine.

---

[9] *Counterintelligence activity* is defined as

> a special kind of activity to ensure state security which is carried out with the use of a system of counterintelligence, search, security and administrative-legal measures and is directed at the prevention, timely identification and repulsion of external and internal threats to the security of Ukraine, of intelligence, terrorist and other illegal actions of special services of foreign states, organisations, groups and persons against the interests of Ukraine.

See Law of Ukraine, "On Counterintelligence Activity," Article 1, 2005.

[10] The number of personnel in the SBU has shrunk since 2005, when it had 41,750 people. See Oleksii Petrov, "Political and Budgetary Oversight of the Ukrainian Intelligence Community," thesis, Monterey, Calif.: Naval Postgraduate School, September 2007, p. 66.

While the SBU is perceived as somewhat better financed and more professional than other Ukrainian security services, there are ongoing questions about its effectiveness. Separatist and Russian agents are believed to be present in the security and defense sectors. Concerns about Russian penetration significantly reduce intra- and interagency trust within Ukraine, which in turn makes it difficult for the SBU to lead counterintelligence efforts. Shifting control over military counterintelligence to the MoD, as currently planned, would reduce some interagency friction.

The roles and subordination of the National Guard are uncertain. The National Guard was created in March 2014 out of the Internal Troops of the MIA. The Law on the National Guard established its structure and defined its functions as a military organization within the MIA, capable of supporting both military and policing missions. But the relationships between the National Guard, the MIA, the Armed Forces of Ukraine, and Ukraine's political leadership are unclear. Most critically, there are questions about the chain of command. Although the National Guard is a component of the MIA, its operational command is in question. In the history of the Anti-Terror Operation (ATO), especially in the retreat from Debalt'seve, it appears that the National Guard was not consistently under the same chain of command as the Armed Forces of Ukraine. Questions concerning the National Guard's role in peacetime are also unresolved.[11]

The security situation in Eastern Ukraine has placed heightened demands on the State Border Guard Service. The situation in Ukraine's East and South has resulted in significantly increased demands for stronger border control capabilities, which would be more in keeping with a militarized force. Ukraine is now facing the need to police its borders in both more- and less-peaceful areas and to provide border security in high-threat situations. Although Ukraine

---

[11] In February 2015, shortly after ceasefire negotiations between the Ukrainian government and the Separatists, Ukrainian forces in the town of Debalt'seve were attacked by Separatist forces and forced to retreat, experiencing heavy casualties (Andrew Kramer and David Herszenhorn, "Ukrainian Soldiers' Retreat from Eastern Town Raises Doubts for Truce," *New York Times*, February 18, 2015).

has received support from other countries, including the United States, the State Border Guard Service is not yet able to fully secure Ukraine's long borders, particularly in the contested and Separatist-controlled areas.

Coordination in internal crises and conflicts is problematic. Ukrainian law and practice designates the command responsibilities of the Ukrainian security organizations for different types of internal security contingencies. According to current law, the Armed Forces of Ukraine have the main responsibility for military conflicts (e.g., local and regional wars); the State Border Guard Service for armed conflict on Ukraine's borders and to protect Ukraine against illegal migrants and smuggling; the National Guard for armed conflict inside Ukraine; and the SBU for counterterrorist activity. In practice, the roles remain undefined. Poor coordination is due to several factors:

- Threats do not always fall clearly into these categories.
- The existing capabilities of the responsible organization may be insufficient to carry out its designated tasks.
- Chains of command are insufficiently clear.
- Coordination is insufficiently developed to ensure that organizations work together rather than independently.
- Political constraints can prevent the clear identification of a crisis.

The ATO has exposed gaps in coordination between the regular army, National Guard, and other government agencies. The SBU is nominally in charge of the ATO, but the GS is meant to play the critical command-and-control role. Government and nongovernmental forces often act independently instead of coordinating efforts. Volunteer battalions operate separately. National and local coordination centers exist, but remain ineffective. Ukraine does not appear to engage in sufficient planning or exercises that could ensure a smooth handoff between different responsible security organizations as a crisis escalates.

Recommendations

Continue to pursue the MIA's reform plan of making itself a small supervisory organization responsible for directing and coordinating a range of separate agencies. The CoM would have overall responsibility for coordinating and directing these agencies through the MIA.

Define the SBU as a domestic intelligence organization, with more clearly and more narrowly defined authorities.[12] The SBU would remain responsible for gathering intelligence on threats to Ukraine within the state, including responding to counterintelligence and counterterrorism, with these terms more narrowly and specifically defined.[13] The SBU could retain responsibility for some law enforcement activities related to internal conflict and crises, such as investigations of acts of terrorism, or these could be transferred to other law enforcement agencies with an understanding of the need for close coordination with the SBU. The Armed Forces of Ukraine and the National Guard should assume responsibility for military counterintelligence.

Coordination and sharing between the SBU and law enforcement and security organizations should be mandated and facilitated through institutional change. Subordinating the SBU to the MIA and the CoM (as in France, Germany, and Poland) would improve coordination, although it may not be feasible in the short term due to the conditions posed by current security requirements.

The SBU should become a recognized agency within Ukraine's intelligence community, as its tasks are largely intelligence functions. Explicit recognition will facilitate closer coordination with the Foreign

---

[12] For example, the British MI-5's mandate is "the protection of national security and, in particular, its protection against threats from espionage, terrorism and sabotage, from the activities of agents of foreign powers and from actions intended to overthrow or undermine parliamentary democracy by political, industrial or violent means." Law of the United Kingdom, "Security Service Act of 1989," Chapter 5, Section 1, 1989.

[13] *Counterintelligence* is defined here as catching spies and safeguarding clandestine, military, and diplomatic operations, and *counterterrorism* is defined as deterring or responding to acts of terrorism (i.e., violence by nonstate actors against civilians intended to instill fear and coerce governments or societies).

Intelligence Service about Russian activities, as well as with the rest of Ukraine's intelligence community.

The size of the SBU should be reduced to increase its efficiency, reduce its cost, and help ensure adherence to democratic norms. This would involve transferring functions to other organizations, contracting out support functions, and focusing only on priority tasks.

Organize the National Guard as a separate agency under the MIA to act as a European-style gendarmerie. Following models of the gendarmerie forces in France, Spain, Italy, and elsewhere in Europe,[14] the National Guard during peacetime would support the local and regional police and the State Boarder Guard in maintaining domestic security and upholding the rule of law, including by providing the ability to respond to challenges to the public order. During war or internal conflicts in which the Armed Forces of Ukraine are in command, including the ATO, National Guard units should be fully subordinate to the military command under the Joint Operational Command (JOC). While the MIA would remain responsible for administrative support, MoD logistics would provide basic supplies such as fuel, food, and ammunition. The Armed Forces of Ukraine sectoral or brigade commanders would have command authority over National Guard troops operating within their areas of operation. This would ensure a line of military command through the Minister of Defense to the President. In internal crises where Ukraine does not face a military threat, the MIA, National Guard, or local officials should have primary authority, depending on the situation. In peacetime, the National Guard should act in support of the police and other organizations within the MIA.

Maintain the State Border Guard Service as a nonmilitarized law enforcement body under the MIA, supported by the National Guard (and the Armed Forces of Ukraine when needed) in times or areas of high threat. The State Border Guard Service would assist the military border units but not engage in combat activities. This approach reflects common NATO and European practices

---

[14] Derek Lutterbeck, *The Paradox of Gendarmeries: Between Expansion, Demilitarization and Dissolution*, The Geneva Centre for the Democratic Control of Armed Forces, 2013.

and conforms to global norms for avoiding the use of armed forces for civilian law enforcement. We also recommend that Ukraine consider merging the Customs Administration with the State Border Guard Service.

Create a clear chain of command for these organizations in internal crises, both at the central government level and in the field, and set up and practice arrangements for coordination. Organizations with relevant capabilities need to be given the lead, ideally with a single point of authority to ensure an integrated strategy that avoids duplication. Coordinating arrangements need to be set up, both centrally and at the local level, and a physical operational center can be highly beneficial for exercising command, exchanging information, and coordinating activities.

CHAPTER THREE

Organizing for Defense

In this chapter, we provide recommendations for a comprehensive new organizational structure for Ukraine's defense institutions, aligned with NATO standards and practices.

Structure of Defense Institutions

What Does the System Need to Do?

There is no official "NATO standard" defense organization; there is considerable diversity among the 28 NATO nations in how they structure their defense institutions. Nonetheless, there are broad commonalities, shared values, and outlooks among the NATO allies. Civilian control of military forces, for example, is a principle embraced across NATO. Other principles include accountability, transparency, strong safeguards against corruption, and public trust in the integrity of the state's institutions that cannot be manipulated by the whims or weaknesses of particular political leaders. Training and development of forces is almost always separated from operational commands, and most NATO defense organizations have an operational command under the Chief of Defense Force (CHoD). The organizational scheme of "J-codes" (i.e., J-1 through J-8/9) is also widely employed. There are other broad, although not universal, commonalities, such as a general preference for professional militaries over conscript forces.

Based on the literature and experience of defense sector reform and effective defense practices globally, one can identify key Euro-

Atlantic standards that guide most effective modern defense organizations today:

- clearly defined roles, responsibilities, and authorities
- clear authorities and accountability for individual performance and action
- delegation to the lowest-possible level to keep senior leaders focused on the most-important topics
- effective civilian oversight
- separation of some key functions, such as setting requirements and procurement; force generation; and operational military command
- cooperation and collaboration on shared tasks through committees and boards for development of requirements, policies, and strategies, and for financial management.

Problems in Ukraine's Current System

The division of roles and responsibilities between the MoD and the GS has changed a number of times since Ukraine became independent. See Figure 3.1 for the defense organizational structure of Ukraine as of fall 2015. The Minister of Defense and the CHoD report directly to the President.[1] There are five Deputy Defense Ministers reporting to the Minister of Defense and six Deputy Chiefs of the GS reporting to the CHoD.

Confusion, gaps, and overlaps in MoD and GS functions preclude effectiveness and limit oversight. Civilian control remains weak in Ukraine except at the very top (that is, in the offices of the President and Prime Minister). Former and current uniformed military personnel dominate the MoD; there is no cadre of professional civilians to run it. The CHoD reports directly to the President concerning military operations, which tends to disempower the Minister in this critical area.

Weak civilian control translates into a system that is more likely to adopt a military answer for security problems than to consider the

---

[1] We use Chief of the Defense Force as the title for the head of the General Staff. The position in Ukraine today is called Chief of the General Staff.

Figure 3.1
Current Structure of Ukraine Ministry of Defense and General Staff

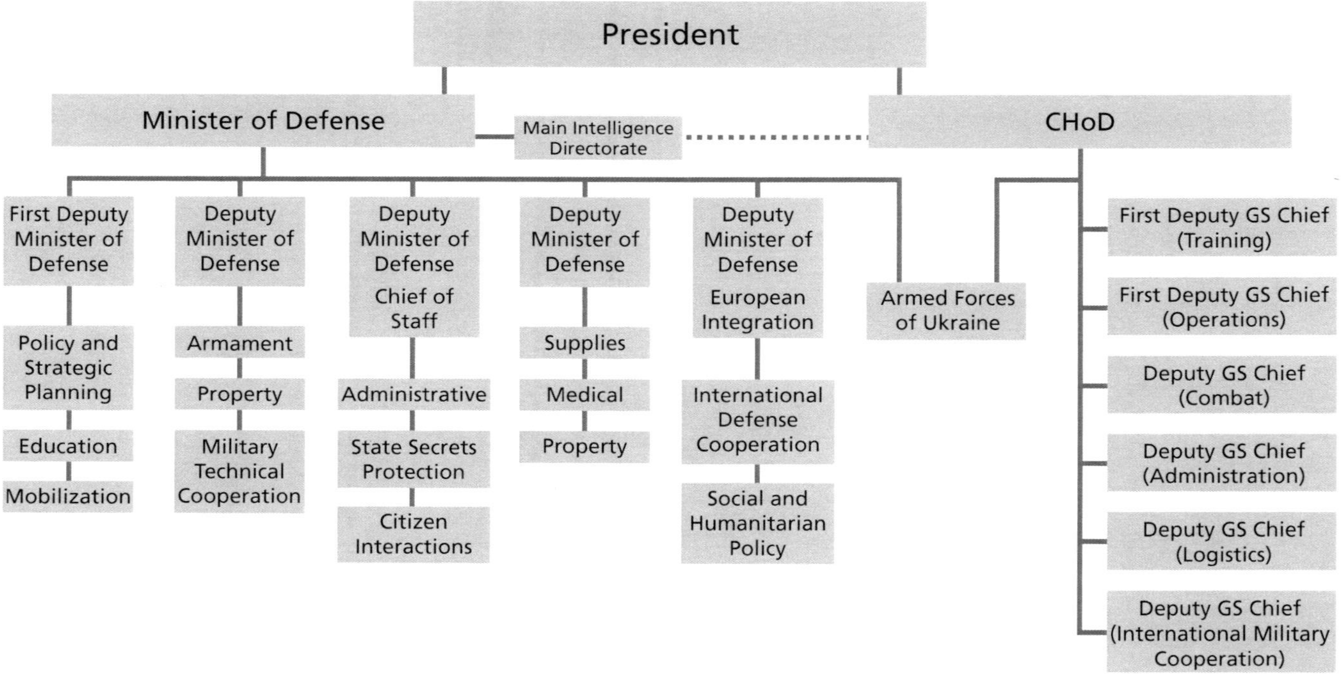

RAND *RR1475/1-3.1*

full range of policy tools available to Ukraine. Divided lines of command and control generate confusion and lead to gaps and overlaps in responsibilities, as well as communication failures.

We have identified a number of areas where the current MoD/GS organization and division of roles and responsibilities hamper effective fulfillment of the missions of the MoD and GS.

- The Minister of Defense is insufficiently empowered. The CHoD is able to make policy-level decisions without the Minister's input.
- The division of roles and responsibilities between the MoD and GS limits both responsibility and effectiveness rather than promoting them. This is because there are both overlaps and gaps between these organizations' functions, and there is no clear division between policymaking on the one hand and policy execution on the other. The lack of clear authority begins at the top of both organizations: Neither the Minister nor the CHoD is ultimately in charge, precluding organizational effectiveness and civilian control. The only official with the authority to coordinate and ensure integration between the MoD and GS appears to be the President.
- The current division of tasks between the MoD and GS creates challenges for identifying responsibility for problems and integrating lines of effort. For example, in the area of procurement, the GS is responsible for requirements and the MoD for purchasing, which makes it difficult to attribute responsibility when purchased items fail to satisfy needs. According to our discussions, the division of military intelligence functions and policies between the MoD and GS is confusing even to staff members themselves.
- The CHoD has both administrative and operational leadership roles—this individual is in charge of both generating forces and commanding operations. Many countries (e.g., Australia and many countries in NATO) divide these functions because the tasks are fundamentally different and the span of control is too large for one individual.
- Regulations and organizational culture tend to push decisions upward to at least the Deputy Minister level, if not to the Minister. Anecdotes abound of Ministers facing up to 500 documents

each day that require their signature, even for matters that could easily be handled four or five echelons lower.

- This mode of operations slows day-to-day decisions and makes it difficult for higher-level individuals to focus on key organizational and operational reforms. Culturally, the result is avoidance of responsibility.

Recommendations: Minister of Defense

We have developed a new defense institutional structure for Ukraine that is economical, effective, and accountable and that closely follows Euro-Atlantic standards. See Figure 3.2 for our proposed organizational structure for the Ukrainian MoD.

Minister of Defense

In our recommended defense organization, the Minister of Defense is the senior official charged with making and carrying out the Ukrainian government's policy on defense. The Minister is the senior adviser to the President, Prime Minister, and the Verkhovna Rada, and is the principal staff assistant to the President on defense policies. The Minister represents the defense sector in the CoM and on the NSDC, and communicates defense resource needs to the President, Prime Minister, CoM, and Ministry of Finance. The Minister also represents the defense sector in interagency deliberations and with international counterparts.

The Minister is responsible for administering the Armed Forces of Ukraine, as well as for military command and control, including command of all forces deployed for military contingencies in the territory of Ukraine. The GS and Armed Forces of Ukraine are subordinate to the Minister.

The Minister would be nominated by the President and approved by the Verkhovna Rada by a majority vote. The President would be able to dismiss the Minister with the approval of the Verkhovna Rada. To ensure strong civilian control, the Minister would be a civilian. If a prospective appointee has military experience, we recommend that Ukraine require that the individual have left military service a minimum of ten years prior to serving in this position. This period of time will ensure the Minister's independence from the military and, more

Figure 3.2
Proposed Organizational Structure for the Ukraine Ministry of Defense

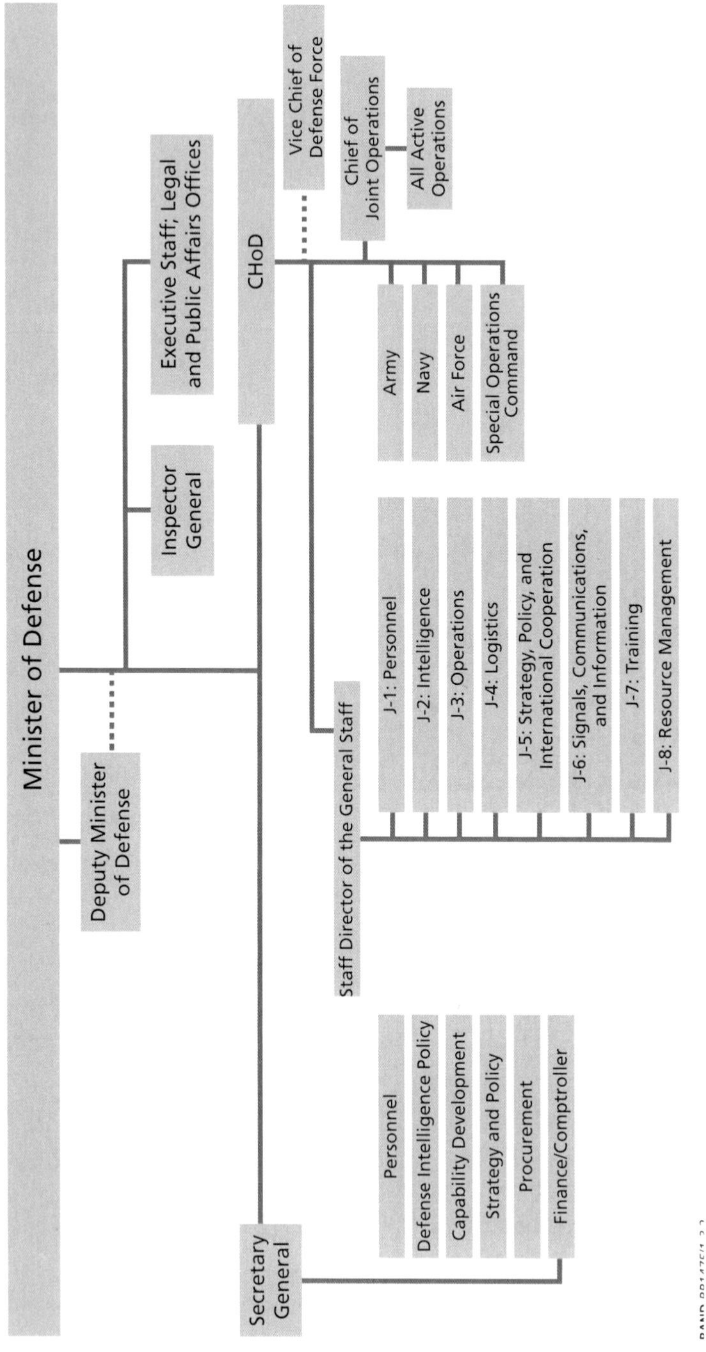

specifically, from former armed service colleagues. (However, in light of the current security situation facing Ukraine, the sitting Minister may remain in his current position.) While a background in defense matters might be useful for a Minister, it would not be a requirement. Demonstrated leadership skills and a strong relationship with the President and the Verkhovna Rada are more important.

If the MoD is to function more effectively, the Minister of Defense will need to delegate authority for decisions to appropriate offices and organizations beneath him or her. Each organization within the Ministry should be empowered to make decisions and take actions within its area of competence and responsibility. While specific roles and responsibilities are and should be specified through laws, regulations, decrees, and directives, they cannot be straitjackets: Decisions should be made at the lowest level possible, and individuals should be held accountable for decisions in their areas of responsibility.

Organizations Reporting Directly to the Minister of Defense

The Minister should have appropriate staff support, including a military adviser and policy advisers on key topics. A number of organizations currently, and appropriately, report directly to the office of the Minister, including the Minister's personal staff, legal affairs, and public affairs. The Inspector General also reports directly to the Minister.

These organizations have reporting lines direct to the Minister because their functions either relate directly to the Minister's representation of the MoD to the wider world and are often highly time-sensitive (Public Affairs), or require immediate access to the senior leadership of the Ministry or the nation in order to maintain independence (Inspector General and Legal Office). Only organizations with such characteristics should have a direct reporting relationship to the Minister. Other organizations (e.g., Comptroller) should fall under the general MoD management system. It is critical not to "re-create" the MoD inside the Minister's office or to create duplicative functions for those elsewhere in the MoD.

The Legal Office, led by the General Counsel, would serve as the senior legal adviser and the chief "lawyer" of the Ministry and be staffed by both civilian and military lawyers. The General Counsel would be appointed by and serve at the pleasure of the Minister. The head of the Office of Public Affairs would be the principal staff assistant to the Minister for public information and be appointed by the Minister. The Office of Public Affairs would manage all public outreach, including the MoD's website and publications.

The Office of the Inspector General would be led by the Inspector General, a senior independent official appointed by the Minister and approved by the Verkhovna Rada and fully vetted to ensure no conflicts of interest. The office itself should comprise a military and civilian staff. The responsibility of this office is to conduct investigations and prepare reports to combat waste, fraud, and abuse. Responsibility for such broader areas as the readiness and adequacy of Ukraine's armed forces lies in the appropriate GS and MoD directorates and offices. The audit department should be a separate organization under the office of the Inspector General.

Deputy Minister of Defense

We recommend the creation of a Deputy Minister of Defense and a Secretary General, rather than multiple deputy ministers as in the current Ukrainian organization, to avoid the risk of divisions and lack of coordination among the different MoD functions.

The Deputy Minister of Defense acts as the primary deputy to the Minister. The Deputy Minister's main function is to help the Minister oversee the entire defense sector; if the Minister looks "up and out," the Deputy looks "down and in." This individual will stand in for the Minister at any meetings or sessions for which the Minister is not available. While the Deputy Minister can act in the Minister's stead, the Deputy Minister should not be an obstacle to accessing the Minister; the Deputy Minister's sign-off is not necessary for ministerial decisions. The Deputy Minister will also serve as a counterpart to deputy ministers from other countries and will participate in deputy-level decisionmaking boards and committees in interagency deliberations. The Deputy Minister will also carry out specific tasks or special projects (e.g., reviews) as directed by the Minister and could oversee certain functions of the MoD and GS, if so delegated by the Minister.

The Deputy Minister should be the second-most-senior civilian in the Ukrainian defense sector, and considered the equivalent of a four-star officer in rank. This individual would be appointed by the President and approved by the CoM. The Deputy can be dismissed by the President with the approval of the CoM. If a prospective appointee has military experience, we recommend that Ukraine require that the individual have left military service a minimum of ten years prior to serving in this position, for the same reasons described in the case of the Minister.

Secretary General

The Secretary General is the principal staff assistant to the Minister and Deputy Minister on all policy matters. The person in this position oversees defense policy in a number of critical areas and ensures that policy reflects the national security aims and strategy of Ukraine's political leaders. The Secretary General provides overall policy advice to the Minister and should work closely with the Ministry of Finance to ensure that the Armed Forces of Ukraine are appropriately and efficiently funded and that those funds are properly allocated.

After the Minister and Deputy Minister, the Secretary General is the most-senior civilian in the Ministry. This individual would be nominated by the Minister and approved by the CoM, although the individual should serve at the pleasure of the Minister, who would be able to dismiss this individual without Cabinet approval.

Individuals/Organizations Reporting Through the Secretary General

The Secretary General enables military capability through the supervision and coordination of policy through six departments: Personnel, Defense Intelligence Policy, Capability Development, Strategy and Policy, Procurement, and Finance/Comptroller. For each of these, the Secretary General assigns a department chief, who would be selected and nominated by the Secretary General, and approved by the Minister of Defense, who can also dismiss them. They may be political appointees or civil servants already serving in the Ministry. All of the chiefs would be civilians; if individuals are former military, they must have left military service a minimum of two years before beginning tenure as department chiefs.

Each department chief would have a deputy to whom a variety of functions may be delegated, as appropriate and agreed upon, and who would have full authority to act on behalf of department chiefs in their absences. Deputies would be civil servants in order to promote continuity and develop the capability of the civil service at senior levels.

Each department would be staffed by knowledgeable and competent specialists, drawn from the civil service or the ranks of active-duty military officers from each service, with the goal that civil servants eventually compose the majority of staff. The department chief and deputy can designate members of the staff to fulfill a variety of functions and to act and make decisions on their behalf when necessary.

When MoD leaders from the Minister on down are unable to make time-sensitive decisions or take action within their area of responsibility for any reason, they would be required to delegate authority; absent another delegation, this responsibility would default to the Deputy or next most-senior person in the chain of command.

Chief of Personnel

The Chief of Personnel would be the principal staff assistant and adviser to the Minister of Defense for management of all personnel in the defense sector, both civilian and military. This position's area of responsibility would include mobilization policy, pay scales, health care policy, civilian and military personnel policy and standards, readiness and training, personnel requirements, and quality-of-life matters affecting all members of the defense sector. This individual would also set the policies that guide human resource decisions for civilians and military personnel assigned to the MoD. Personnel department leadership and staff coordinate with the GS and the armed services on personnel matters and provide policy oversight of functions managed by those organizations. They would participate in planning, programming, and budgeting activities as they relate to personnel matters. In support of these missions, the Chief of Personnel would:

- issue specific policy guidance concerning the overall numbers of personnel—civilian and military—in the defense sector of

Ukraine, as well as personnel allocations for specific organizations, as appropriate
- provide policy guidance on the capabilities and readiness of all personnel in the defense sector
- set pay scales for both civilian and military personnel in the MoD and the Armed Forces of Ukraine
- establish and issue guidance for the entire defense sector on human resources management processes and procedures
- participate in defining the personnel requirements (numbers, training, and readiness) that will be needed to fit the plans and programs put forward for weapon systems by the Chief of Capability Development and the military services
- establish and maintain databases and information systems to track individuals in the system and ensure the accuracy, completeness, and timeliness of these data, and use the most-effective and most-efficient modern technologies and practices.

Chief of Defense Intelligence Policy

The Chief of Defense Intelligence Policy would provide policy advice to the Secretary General and the Minister on defense intelligence matters and security. To identify defense intelligence requirements, the Defense Intelligence Policy Department would work with the intelligence organization in the GS, the military services, and the JOC and provides policy oversight to and coordination for those bodies. While these other organizations maintain responsibility for operational and analytical intelligence tasks, the Department of Defense Intelligence Policy defines the policies that guide those operations and analyses. It would also represent defense intelligence in the interagency intelligence community, although the GS intelligence organization could also participate. The Chief of Defense Intelligence Policy would work with the Department of Capability Development on policy for developing defense intelligence capabilities, with the Department of Personnel on defining policy for recruiting and retaining defense intelligence personnel, and with the Department of Finance/Comptroller on defense intelligence funding needs. Unlike the current Ukrainian Main Intelligence Directorate, this department would not have an independent intelligence collection, operations, or analysis function. These functions would reside in the GS. In support of these missions, the Chief of Defense Intelligence Policy would oversee the development, management, and coordination of intelligence relationships between defense intelligence and the Chief's foreign and international partners and counterparts, and would advise and assist the Chief of Capability Development and the Chief of Procurement on acquisition programs that affect intelligence capabilities and programs.

Chief of Capability Development

The Chief of Capability Development would be responsible for defining the overall requirements for an operationally effective and cost-efficient mix of military capabilities (equipment, personnel, training, infrastructure, etc.) that can achieve Ukraine's strategic objectives. To do this, this department would work closely with the Department of Strategy and Policy, the Department of Finance/Comptroller, the GS, Armed Forces of Ukraine, and the JOC to translate Ukraine's strategic objectives into overall military requirements. It would then work with the Department of Procurement to ensure that these requirements are translated into effective systems and with the Department of Personnel to ensure that the requirements for systems are effectively translated into personnel needs.

This department would be responsible for answering the questions: "How will we translate our strategy into people and systems that can carry out the strategy?" and "What is the optimal force mix to achieve objectives within cost constraints?" As plans are developed and implemented, the office would assess whether these plans would, in fact, meet requirements. While all departments under the Secretary General would be composed of both civilian and military staff, this one would be unique in that it would always include members from every service of the Armed Forces of Ukraine (Army, Navy, Air Force, and Special Operations Command). It would also be unique in that it could be led by either a military officer or a civilian (if military, a civil servant should serve as the deputy; if civilian, a military officer should be the deputy).

Chief of Strategy and Policy

The Chief of Strategy and Policy would be the principal adviser to the Secretary General and the Minister of Defense on all questions concerning defense strategy, planning, and projections of the future strategic environment. This individual would serve as the MoD lead for interagency strategic integration and would develop planning scenarios and assumptions for a potential range of short-, medium-, and long-term conflicts and crises. The Strategy and Policy Department would formulate, coordinate, and distribute strategic guidance statements and would write, coordinate, and issue key defense documents, such as Strategic Defense Reviews. The department would work closely with and provide policy guidance to the work of the GS. This department's approval of budget, procurement, and force-building plans would be required to ensure they align with overall strategy. In support of these missions, the Chief of Strategy and Policy would

- work closely with the Chief of Capability Development to ensure that military capabilities are aligned with and able to carry out the defense and military strategies
- serve as the key point of contact and oversee Ukraine's Euro-Atlantic integration after absorbing the current integration office
- provide advice, insight, and analytical support to the Minister of Defense upon request or keep senior leadership informed of current events
- issue biannual white papers on the Armed Forces of Ukraine, which shall include cost figures as well as force plans
- commission and oversee the conduct of outside studies on strategic matters.

Chief of Procurement

The Chief of Procurement would set policy for and supervise the procurement of weapons and equipment for the Armed Forces of Ukraine. This department would work closely with the Department of Capability Development to translate the broad requirements developed by that department into specific requirements for systems. This organization would also ensure that budgetary decisions take into account the entire life-cycle costs of defense systems, and lead or participate in major defense decisionmaking boards and committees, including cochairing key committees on requirements and standards with the Department of Capability Development. A division of this department would oversee testing and evaluation of materiel procured by the defense sector. The Chief of Procurement would be responsible for maximizing the transparency and efficiency of procurement, while reducing corruption. This individual would oversee and organize program offices to undertake multiyear procurement activities and would recruit and maintain a highly professional procurement workforce, trained in contracting and government regulation. The Procurement Department would also be responsible for management of defense property and infrastructure.

Chief of Finance/Comptroller

The Chief of Finance/Comptroller would be the principal staff officer to the Secretary General and the Minister on all financial and budgetary matters in the MoD. This individual would focus on budgetary formulation and execution; financial management and oversight; financial information, preparedness, and transparency for audit; and accounting policy and procedure. The Chief of Finance/Comptroller would be the chief financial adviser to the Minister and would formulate the overall defense budget. In carrying out these tasks, the Chief of Finance/Comptroller and the staff of the Finance/Comptroller Department would

- lead the development of the annual budget, working closely with the Chief of Capability Development and the Chief of Procurement
- conduct interactions with the staffs of the presidential administration, the NSDC, and the Ministry of Finance on budgetary and fiscal matters, and the execution and control of budgets
- maintain effective control and accountability over the use of all financial resources of the MoD
- conduct analyses aimed at increasing the efficiency of defense spending in coordination with the Chief of Procurement

- develop and maintain integrated MoD accounting and financial management systems, including financial reporting and management controls
- direct, manage, and provide oversight of MoD financial management personnel, activities, and operations.

Recommendations: Chief of Defense Force

The CHoD would be the senior military officer in the country and responsible for leading the GS and the Armed Forces of Ukraine. The military operational chain of command would run from the President to the Minister of Defense, to the CHoD, to the Joint Operational Commander, and on down through the military units. The CHoD would be the primary military adviser to the Minister and to the President and would be able to advise the President directly on operational matters, but would have to keep the Minister of Defense fully informed. There needs to be a collaborative relationship between the CHoD and the Minister of Defense.

The CHoD would lead the planning and conduct of military operations, delegating day-to-day responsibility for operations to the Chief of Joint Operations. The CHoD would communicate the resource needs of the Armed Forces of Ukraine through the Minister to the CoM and the President. The CHoD would delegate responsibility to man, train, and equip the Armed Forces of Ukraine to the GS and the individual military services. The CHoD would be responsible for the overall condition of the Armed Forces of Ukraine, including their capabilities and readiness, according to standards specified by the Minister and the Secretary General. The CHoD would also help build Ukraine's contacts with Euro-Atlantic and other international military forces.

The CHoD would hold a four-star rank and be appointed by the President to a three-year term, which could be extended by a year for a potential total tenure of four years. The CHoD would serve at the pleasure of the President and Minister of Defense and could be dismissed as a result of failure to adequately perform the duties of the office.

Individuals and Organizations Under the Chief of Defense Force

In leading the Armed Forces of Ukraine, the GS, and the Chief of Joint Operations, these organizations and individuals report through the CHoD to the Minister of Defense.

Vice Chief of Defense Force

The Vice CHoD would act as the primary deputy to the CHoD. This person would stand in for the CHoD at any meetings or sessions for which the CHoD is not available, serve as a counterpart to international Vice or Deputy CHoDs, and participate in deputy-level decisionmaking boards and committees. The Vice CHoD would also carry out specific tasks or special projects as directed by the CHoD.

This position would be the second-most-senior officer in the Armed Forces of Ukraine, with a three-star rank. The President would appoint the Vice CHoD to a three-year term, which could be extended by one year for a potential total tenure of four years. Like the CHoD, the Vice CHoD would serve at the pleasure of the President and the Minister of Defense and could be dismissed as a result of failure to adequately perform the duties of the office. The Vice CHoD would need to maintain a strong relationship and communications with the Minister, Secretary General, military services, and chiefs of departments.

Staff Director of the General Staff

The Staff Director of the General Staff would act as a deputy to the CHoD with specific responsibility for managing the GS. This individual supervises the work of the GS in providing advice and assistance to the CHoD, the Vice CHoD, and the Minister of Defense. The Staff Director will coordinate with the Secretary General and facilitate coordination with the chiefs of departments under the Secretary General.

The Staff Director would be the third-most-senior officer in the Armed Forces of Ukraine and equivalent in rank to the chiefs of the individual military services. This individual would be appointed by the CHoD for a three-year tenure.

General Staff

The GS would be the primary staff of the CHoD and Vice CHoD. To align with Euro-Atlantic standards, the GS should be reorganized into traditional J-code functions, which would also carry through to the services, JOC, fighting units, and throughout the force:

- J-1: Personnel
- J-2: Intelligence
- J-3: Operations
- J-4: Logistics
- J-5: Strategy, Policy, and International Cooperation
- J-6: Signals, Communications, and Information
- J-7: Training
- J-8: Resource Management.

The GS would participate in deliberations and decisionmaking committees that define requirements, procurement, personnel, and budget allocations. To do this, the GS would work closely and consult with the relevant departments under the Secretary General, the individual armed services, the Chief of Joint Operations, and the JOC.

The GS would be composed mostly of military officers from all military services but would also incorporate some civilian experts, as feasible and appropriate.

Chief of Joint Operations

The Chief of Joint Operations would be the primary operational military commander and provide command and control for all current operations and exercises undertaken by the Armed Forces of Ukraine, each of which would have its own commander subordinate to the Chief of Joint Operations. This position would also be responsible for developing current and future operational plans in consultation with other MoD components, such as the Chief of Strategy and Policy, the Chief of Capability Development, and the GS. This individual would also be the commander of the JOC, which features an operational staff that is also organized according to J-codes. This staff would focus exclusively on the conduct and support of ongoing and potential future operations as directed.

The Chief of Joint Operations would also communicate requirements and current needs for ongoing operations, primarily through the CHoD. The person in this position would be the commander of subordinate leaders of current operations and could remove them from their posts and replace them, subject to consultation with the CHoD.

The Chief of Joint Operations would hold the rank of a three-star general or admiral and be subordinate to the Minister, the CHoD, and the Vice CHoD. This individual would report to the Minister of Defense, through the CHoD. The Chief of Joint Operations would be selected by the CHoD, in consultation with the Minister and with the approval of the President, and assigned a three-year tenure. This individual would serve at the pleasure of the President and Minister of Defense and could be dismissed for nonperformance.

Implementation of Defense Structural Recommendations

Successful implementation of structural reforms needs to be guided by a set of principles that can be translated into specific actions by Ukraine's government. Implementation will need to be tailored to the specific circumstances and institutional culture of Ukraine, however.

The first principle is that reform starts at the top. Senior officials across the Ukrainian government will need to commit to the reforms and work together to implement them through the various new organizations.

The second principle is that a single person must be responsible for implementing the reforms. That person should be the Minister of Defense, given the responsibilities assigned to that position in the structural reforms recommended in this report. The Minister should be responsible for making final decisions on the new organizational structure, identifying individuals and delegating to them the day-to-day implementation of the reforms, and ensuring that the reforms are implemented. In carrying out this responsibility, the Minister would be accountable to the President.

Once given the responsibility for implementing the reforms, the Minister of Defense will need to ensure that the various stakeholders in the MoD and GS are involved as the reform is implemented. This will need to happen not only through formal groups, but also through informal interactions.

The Minister of Defense will need staff support to implement and coordinate the reorganization and reform. We recommend that the Minister appoint a full-time "Director of MoD Reform Implementation" who has stature within the government and the trust of the major stakeholders. This person would put together a task force composed of dedicated staff, drawing on individuals from different parts of the MoD and GS. This task force would produce working papers detailing different aspects of implementation and, where necessary, make recommendations for further decisions needed on the part of the Minister or other senior leaders to implement the reforms.

We also recommend that the Minister of Defense set up and chair an "Advisory Group on Implementation." This group would include the CHoD as well as individuals who will be appointed to leadership positions in the new organization. This advisory group should create subgroups for different aspects of the reform (e.g., one for the MoD, and one for the GS). Informal and ad hoc groups of key stakeholders could also be pulled together as particular matters arise. This group will need to work with the Director of MoD Reform Implementation to identify which departments need to be shifted from the GS to the MoD or from the MoD to the GS. It will also need to approve recommendations by the Director of MoD Reform Implementation concerning reductions in staff. Decisions on who will head the new departments will need to be made by the MoD or GS.

Senior officials to lead the new organizations within the reformed structure will need to be chosen quickly. The Deputy Minister of Defense, the Secretary General, and the department chiefs will need to be appointed for the MoD; the Vice CHoD and Staff Director for the GS. These individuals would be given responsibility for implementing the reforms within their new organizations and would work closely with the Director of MoD Reform Implementation. They will also become members of the Advisory Group on Implementation.

The Director of MoD Reform Implementation and task force will need to identify those reforms that will require changes in Ukrainian laws. For example, the proposed changes in the roles and responsibilities of the Minister of Defense will necessitate changes in the "Law on the Armed Forces of Ukraine" and the "Law on Defense of Ukraine" as well as several cabinet resolutions and presidential decrees. The Verkhovna Rada will need to be involved in any changes in laws, while the President and the CoM will need to be involved in changing resolutions and decrees. The Director of MoD Reform Implementation will need to set up a legal task force to identify all laws, resolutions, and decrees that will be affected by the proposed reforms and draft the language needed so that the proposed reforms are consistent with Ukraine's legal system.

Internal regulations will need to be revised inside the MoD and GS. Changes in personnel policies, in particular, will need revisions, including job classifications and descriptions, benefits, salary schedules, performance reviews, and procedures for promotions. These personnel changes will need to be done as quickly as possible to attract individuals to the new positions. Additionally, financial incentives will be needed to enlist those taking on the new jobs and responsibilities—as well as for those whose positions are being eliminated (in the form of buyouts, layoffs, or retirements).

The Minister of Defense will need to set a timeline for implementing the reorganization, setting a date for it to begin (e.g., when laws have been revised) and a date for its full completion, with the possibility of setting different dates within the overall timeline for specific reforms. The Minister will need to balance an appreciation of how long it will take to fill the new positions with a desire to move as quickly as possible so as to reduce the bureaucratic resistance that will arise.

CHAPTER FOUR

Defense Reforms to Improve Warfighting and Efficient Use of Resources

In this chapter, we turn to four critical areas that have particular importance for Ukraine's warfighting and efficient use of resources. These areas (among others) have also been identified by NATO as priority areas for assistance through the trust funds.[1]

Without the ability to move information, orders, and other communications from the highest levels of government through the intermediate commands and to the front lines through effective command, control, communications, and intelligence (C3I), senior leaders will not be able to convey their orders, combat forces will not have the intelligence they need to operate, and military units will not be able to coordinate their activities. Personnel must be motivated to fight and have the proper training to be effective. Recruitment and retention policies must be financially sustainable to avoid wasting Ukraine's limited resources and to ensure necessary manning of the force. Combat forces need modern, effective, and quality weaponry and equipment, and Ukraine needs to be able to procure items within its limited budgetary resources. While Ukraine is a major arms producer, efficient procurement for ongoing military operations was not a major priority prior to 2014. Without sufficient quantity and quality of food, fuel, and ammunition, militaries cannot operate. Weapons, equipment, and supplies need to be efficiently tracked to reduce loss and ensure rapid resupply.

---

[1] NATO, "NATO's Practical Support to Ukraine," Fact Sheet, June 2015.

Command, Control, Communications, and Intelligence

What Does the System Need to Do?

Both in the current conflict and for the long term, Ukraine's C3I system needs to provide

- a clear chain of command starting at the top, with roles and responsibilities of each echelon clearly specified in writing, and authority appropriately delegated to subordinate commanders
- for each operational commander, particularly commanders at the ATO command center and sector-level commands: real-time information about what is happening around their battle space; competent operations planners; rapid, secure communications with appropriate subordinate echelons of command; and the flexibility to maneuver based on fast-moving events
- for tactical commanders: operational warning of enemy actions outside the coverage of the tactical commander's own resources; secure communications to subordinate units, as well as to leadership and other commanders under conditions of radio-electronic warfare; and regular procedures and means for coordination with adjacent units.

Policies, practices, and equipment accepted and incorporated during the current conflict in the ATO can become the foundation for effective C3I incorporated into the long-term design of the armed forces.

Problems in Ukraine's Current System

Command and Control

The current conflict has been designated an ATO, which puts the SBU, rather than the MoD or GS, in the formal lead of the operations, making it difficult at times to achieve unity of command.

The current Central Command Center is not functioning as an effective operational command. The elimination of the JOC in 2010 removed a critical operational command echelon.

Senior officials have at times reportedly sidestepped the formal chain of command through direct communications to brigades or

even battalions. These practices have confused commanders in the field and introduced uncertainty regarding delegation of responsibility.

Responsibility for current operations is not centralized—operational and tactical commanders have insufficient authority and responsibility. The current state of affairs requires intervention from senior commands, implying that senior commanders, who are responsible for other critical tasks, must also manage operations.

Communications

Outdated equipment and continuing lack of secure communications hampers military effectiveness. Ukrainian communications can be easily intercepted by the adversary, compromising security and undermining operations. Volunteers are supplying some secure and insecure communications equipment, including insecure Motorola radios, for use within brigades. There are shortages of equipment and the systems are not always compatible.[2] This is particularly challenging when different types of forces (e.g., Armed Forces and National Guard) need to communicate. Automation efforts in logistics and personnel are held back by insufficient capacity for secure communications. For example, an electronic personnel data management system that would enable Ukraine to move past cumbersome paper systems for tracking enlisted personnel is apparently under development, but insufficient secure communications capacity prevents it from being implemented. An effective automated warehouse system that enables visibility across the entire logistics system will similarly require greater bandwidth.

Existing military and security forces do not routinely communicate intelligence and operational information horizontally.

Recommendations

Performance gaps identified in the current system spring from complicated interactions among legacy organizations, personnel, and equipment, combined with the unique situation presented by the ATO. Even the deficiencies specific to the current fight highlight longer-term,

---

[2] There are ongoing efforts by NATO and NATO allies to provide additional secure communications equipment.

broader problems with existing arrangements. Our recommendations are meant to directly address weaknesses revealed by operations in the ATO and introduce longer-term, more-general improvements into the overall system. Specifically, the recommendations include clarifying the chain of command for the ATO. With this as the central change, further recommendations expand on key enablers to the effective functioning of command.

Chain of Command

As we previously noted, we recommend the appointment of a Chief of Joint Operations as the single point of contact to give direction to operational commanders. The chain of command would run from the President to the Minister of Defense and through the CHoD to all operating forces (see Figure 4.1). The Chief of Joint Operations receives policy guidance and direction from those above him or her in the chain of command and translates them into formal, authoritative direction as necessary. The Chief of Joint Operations should lead the JOC, which includes all active operations. The JOC should include a J-coded staff that supports the Chief of Joint Operations and interacts with their

Figure 4.1
Proposed Operational Chain of Command

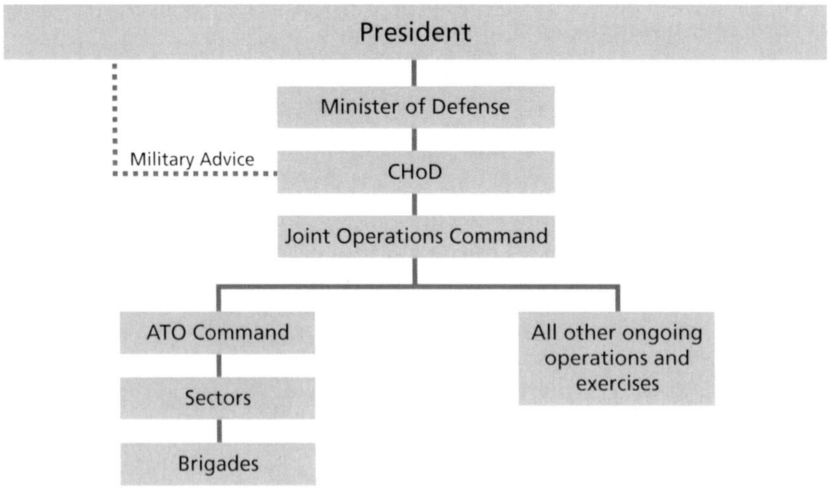

RAND *RR1475/1-4.1*

counterparts on the GS and in the MoD. Creation of the Chief of Joint Operations and JOC would centralize responsibility for ongoing operations, ensuring that a single individual can devote full attention to critical tasks and that the rest of the defense establishment provides for the urgent needs of combat units.

The Chief of Joint Operations should make recommendations to those higher in the chain about allocating forces to operations both initially and as operations develop, and should be the authority subordinating those forces to operational commanders. The Chief of Joint Operations also has responsibility for presenting the resource needs of operational commanders (supplies, ammunition, weapons, personnel and unit replacements, etc.) to supporting agencies and for monitoring JOC J-staff sections' status and performance in ongoing operations.

The JOC should function as a command or headquarters. It should be staffed 24 hours per day and seven days per week to receive and transmit information and intelligence up and down the chain of command and, as appropriate, communicate new assignments and orders to operational commanders. It should also

- provide continuous updates on the status of actions, units, unit strengths, casualties, and logistic states
- respond to inquiries from authorities on any of the above
- maintain the "big board" map and indicators
- function as an appropriate place for more-senior figures to gather during crises and to monitor critical operations
- receive national-level intelligence and disseminate it through J-2 channels to operating commands
- possibly serve as a place to centralize public affairs for the military.

Within the ongoing operations subordinate to the JOC, it is critical to specify responsibilities for each layer, delegate relevant authorities to subordinate officials, and make them accountable for the performance of these tasks. Superior echelons must not routinely skip layers in the chain of command. Superior officers must permit subordinates to take initiative and perhaps fail, or else subordinates will not take responsibility for their own actions.

Within the ATO, the commander should have full authority and operational control (in the NATO sense) over all units and government organizations within the area of operations. While the Chief of Joint Operations should provide guidance on overall strategic objectives and should supervise the performance of the ATO commander, the commander should be in charge of the operation. The sector commanders are the direct subordinates of the ATO commander, and should function as the operational commanders of their sectors and have authority over all elements of the national forces within their areas of operations.

Command Enablers

To be able to command its military forces effectively, Ukraine needs to expand on key enablers. Specifically, we recommend that Ukraine

- conduct exercises to practice new command arrangements to address gaps and deficiencies from past operations and incorporate experiences into planning. Built around possible situations that might arise in the next fighting season, these exercises would enable commanders and their staffs to understand the capabilities of other units, build trust, and accustom themselves to their roles in any future operations. Use after-action reviews to identify skills and personnel gaps.
- incorporate recent lessons into planning, curriculum, and training for future commanders and the preparation of new officers. Integrate lessons identified into training by ensuring that the ongoing lessons-learned effort in the GS includes both a planning and a training component: The process should identify how these lessons should affect planning and training, and the results of this analysis should be communicated to both planners and trainers and integrated into their work. The lessons-learned effort should seek to leverage existing informal commanders' discussion networks to elicit lessons that might not otherwise be captured by formal processes.

Ukraine also needs to focus on improving communications by

- acquiring enough secure communications equipment to support operations in the ATO. This should be a procurement priority as

well as a priority for foreign donors, with appropriate accountability in place to enable continued foreign assistance. Over time, military and National Guard units should adopt interoperable radios and other communications equipment.
- adapting or discarding existing and newly acquired nonstandard equipment as necessary.
- beginning the design of long-term secure data networks and identify those portions of the existing and newly added equipment sets that can be used in that system.
- beginning to monitor Ukrainian communications, including such unofficial means as mobile phone networks. Assign responsibility for the communications-monitoring mission, analyze what intelligence communications lapses provide to the adversary, address shortcomings by remedial instruction and disciplinary measures, and identify lessons for future systems and equipment acquisition.

Finally, we recommend that Ukraine direct and establish the means by which intelligence is passed from the MoD and intelligence organizations down to the units, and from the units up to the top of the organization.

Personnel, Recruiting, and Training

What Does the System Need to Do?

Ukraine's personnel system needs to provide highly trained, educated, and motivated enlisted personnel and commissioned and noncommissioned officers to the armed forces. Military doctrine and training should create and sustain a culture that empowers these individuals to make quick decisions and take creative and effective action. The personnel system needs to recruit qualified personnel by providing competitive compensation, including both wages and benefits, and a clear trajectory for promotion and advancement. It needs to provide appropriate training for those individuals, after which it needs to be able to deploy these individuals to where they are most needed within a reasonable period of time. It also needs to have effective, objective systems for promoting, retaining, and dismissing personnel. Ukraine needs an experienced, able, and effective strategic and operational reserve force, which the personnel system will need to be able to attract and retain. It will also need to provide appropriate benefits to demobilized personnel and veterans. As part of the overall Ukrainian government personnel system, it should also provide trained, experienced, motivated, and loyal civil service personnel to the MoD and GS who are capable of providing policy guidance and expertise on the full range of defense matters.

Problems in Ukraine's Current System

Recruiting and Mobilization

Reliance on conscription and mobilization undermines current and long-term goals. To meet increased troop requirements early in the ATO, Ukraine mobilized personnel with and without prior military service through a Soviet-legacy system. Ukraine reinstituted conscription and mobilization to create an army of mixed professional and conscripted personnel.

The ATO is a high-intensity conflict, requiring highly trained personnel. In this type of conflict, training and the quality of the force are more important than numbers. Conscription and mobilization are not good methods for creating a force capable of fighting in the current conflict. For both the current conflict and most likely contingencies, Ukraine would find an all-volunteer professional force to be more effective.

Ukraine's current system of mobilization is expensive and undermines the development of an all-volunteer force. Mobilized personnel receive their premobilization salary in addition to their military salaries, while "contract" or professional personnel not serving at the front receive relatively low salaries. This system discourages demobilized personnel from signing contracts—they may fear being assigned to units not at the front and therefore paid poorly, and may seek to avoid being locked into a long-term military contract once the war ends. The costs of mobilization also prevent the Ukrainian government from increasing salaries for contract personnel and often lead to funding shortfalls. Moreover, as a consequence of short-term mobilization, the Armed

Forces of Ukraine must continue to train large numbers of soldiers and then send untested personnel to the front.

Neither mobilization nor conscription is universal. Therefore, the systems are inevitably unfair. While there have been some improvements in Ukraine's selection system over the past year and a half, some people, especially those with money or political connections, may still be able to escape mobilization or conscription. It could be possible to mitigate these problems by continuing to improve the selection of recruits and reforming the salary structure. Universal mobilization would be more equitable, and may have other nonwarfighting political advantages, including fostering patriotism and broadening military experience and understanding of military culture among the population. However, effectively implementing universal conscription is very expensive and inefficient, as it requires constantly training many more personnel than are actually useful to the force. Ukraine cannot afford to adopt a costly universal conscription system, as Finland and Israel have done. Ukraine faces a trade-off between military effectiveness (which is more likely with a volunteer force) and broader, but still imperfect, burden-sharing (by continuing the current system of conscription and mobilization or a variation thereof). Little financial benefit is to be found from either approach because both professional forces and conscripted or mobilized forces are expensive, and particularly so, given Ukraine's current approach to mobilization. Therefore, we recommend that Ukraine choose military effectiveness, and thus a volunteer force.

In addition to the challenges of attaining overall goals for force size, Ukraine faces specific challenges in recruiting a sufficient number of junior officers, developing a cadre of noncommissioned officers, and developing a civilian cadre within the MoD.

Compensation System

Ukraine's current salary and benefits structure for both military personnel and civilians creates distorted incentives and does not efficiently attract and retain personnel.

The current structure features low base pay, substantial bonuses for service in the ATO and for possession of special skills, and provides expensive fringe benefits that do not serve to attract the desired mix of recruits. Entry-level salaries for enlisted soldiers are significantly below market rates, with base pay for volunteer soldiers at 2,340 hryvnia per month, according to 2015 data provided by the MoD. At the time, this figure was 20 percent lower than wages for a competing occupation: Construction workers earned an average of 2,950 hryvnia per month at the same point in time.

Despite past promises, Ukraine has long been unable to provide military housing in accordance with current regulations. Many active-duty personnel do not receive service housing. The lack of housing for active-duty officers provides strong disincentives for officers to rotate, effectively bifurcating field and staff officers. Retirees face long queues for the housing they have been promised, and construction of housing by the MoD appears costly and inefficient. The Ukrainian government will need to revamp this system, which proved untenable long ago.

Training

While the MoD and GS are engaging in a gradual process of improving training and education, several important gaps remain.

There appear to have been significant improvements in training since the beginning of the war, including an updated curriculum for enlisted personnel; a longer period of training, including two months of basic and specialized training and a month of cohesion training with units; and efforts to appoint ATO veterans as instructors. Interviews with serving military personnel indicate that they feel these changes have had positive effects on the battlefield, although to our knowledge no comprehensive assessment has been carried out. NATO countries also have increasingly deployed training missions. Nevertheless, some gaps remain:

- The curriculum for officer education and testing needs to be updated, particularly to incorporate recent lessons identified through battlefield experience; e.g., the experience of Debalt'seve.
- Equipment available for training is insufficient, tending not to reflect what is issued to forces in the field. While battlefield needs must come first, the lack of equipment presents real challenges for

ensuring that personnel know how to use the equipment they will use in the field.

- Instructors do not themselves receive sufficient specialized training.

Recommendations

Structure, Roles, and Responsibilities

As noted, we recommend Ukraine create a Chief of Personnel under a newly appointed Secretary General, who will act as the principal adviser to the Minister of Defense on the management of all personnel in the Ukrainian MoD, and who will establish policy and oversee policy implementation on personnel, training, and readiness (see Figure 4.2). The Chief of Personnel will be ultimately responsible for developing human resources policies for civilian personnel. Under the Chief of Personnel should be offices responsible for civilian and military personnel. The latter will work closely with the J-1 and J-7 of the GS to ensure that their work aligns with overall personnel policy. The office should be structured to ensure that there are sufficient and capable staff to oversee policy for the military and, as relevant, civilian recruitment, training, and education; veterans' affairs; health and military medicine; and reserve management.

Ukraine should structure its forces along the lines of modern militaries in terms of the number of personnel at each rank and grade.

Figure 4.2
Proposed Chain of Command for Personnel

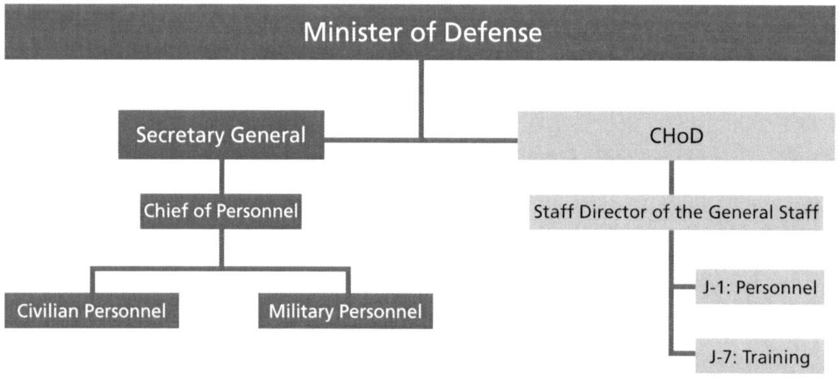

RAND *RR1475/1-4.2*

Table 4.1
Two Potential Force Structures for Ukraine

Rank	Percentage of Force	Total Military Force of 146,000	Total Military Force of 250,000
Civilians		40,000	40,000
Uniformed			
Superior officers (major general, lieutenant general, colonel general—counter-admiral, vice-admiral, admiral)	0.07	69	137
Senior officers (major, lieutenant colonel, colonel—captain [1st–3rd rank], captain-lieutenant)	6.50	6,841	13,552
Junior officers (junior lieutenant, lieutenant, senior lieutenant)	10.20	10,838	21,472
Total officers	16.77	17,748	35,161
Senior enlisted (master sergeant and sergeants-major—midshipman and senior midshipman)	10.40	11,042	21,875
Middle enlisted (junior sergeant, sergeant, senior sergeant, starshina—starshina 1st and 2nd class, chief starshina, chief ship starshina)	52.70	66,672	132,087
Junior enlisted (soldier, senior soldier—matros, senior matros)	19.20	9,578	18,975
Total enlisted	82.30	87,292	172,937

SOURCE: This rank structure was derived from U.S. Department of Defense, "Active Duty Military Personnel by Service by Rank/Grade: April 2015," DoD Personnel, Workforce Reports & Publications, Defense Manpower Data Center, 2015.

NOTE: For each rank, we list Army ranks followed by a dash and Navy ranks. The rank structure was adapted for Ukraine's ranks. It is intended to offer a general guide to a Western rank structure; Ukraine may wish to make some adaptations. Junior enlisted ranks in Ukraine are matched to E-1 to E-3; middle enlisted to E-4 to E-6; senior enlisted to E-7 to E-8; junior officer to O-1 to O-3; senior officer to O-4 to O-6; and superior officer to O-7 to O-9. The figure of 40,000 civilians in the MoD was provided as an approximation of current staff size. A review and subsequent reduction of civilian personnel would be advisable to reduce costs and improve efficiency.

Table 4.1 shows two potential force structures, reflecting two different possible force sizes, which should serve Ukraine's national security needs.

Requirements for personnel, including qualifications for various ranks and positions, content of training, etc., should be developed by committees chaired by the office of the Chief of Personnel, which include representatives from the armed services, the GS, the JOC, and other relevant organizations.

Policy

Ukraine should strive for an all-volunteer force with a reserve component. The active reserve will need to train and exercise regularly, several times annually.

Officers who are no longer contributing should be made to leave the force. We are not recommending, at this time, a policy in which those who are not promoted are encouraged to leave, but we are recommending regular reviews to ensure that all who remain in the force are valuable to it. Such a system will need a steady supply of junior officers to replace older officers who are exiting.

In addition to existing systems for developing junior officers, including university training programs and battlefield promotions plus training, Ukraine should create a four- to six-month officer candidate school for civilian personnel who already hold at least a two-year degree from an institution or college of higher education and seek to join the armed forces.

Ukraine needs a better system for recruiting, training, and retaining professional noncommissioned officers. The MoD should reinstitute the former training academy for noncommissioned officers, advised by Western military personnel.

To avoid losing critical experience, motivated and qualified personnel who have performed well in the ATO should be identified and recommended for future training in a noncommissioned officer academy, officer candidate school, or a military academy. Some personnel are already being identified for promotion or further education on an ad hoc basis, but this process should be formalized.

For the MoD to develop and maintain effective civilian control, even as it reduces overall staff size, it should bring in civilians with relevant analytic and management skills from outside the military system and train them to apply those skills in the area of defense and military affairs. The international community may be willing to provide this training. Immediate efforts should focus on identifying and incorporating capable people who bring useful skills. The ongoing "volunteer" program within the MoD should be leveraged for this effort and made more sustainable with more budgetary support. Fellowship programs, perhaps privately funded, can be an effective mechanism to bring people in for the short term, with the option of converting to civil service, at competitive salaries, at the end of a fellowship term. Another measure should be to identify capable military and civilian personnel with relevant Western training and capabilities and promote them to senior positions. Efforts should be made to substantially increase the proportion of civilian personnel in the future by working with universities on degree programs, scholarships, and internships in the MoD.

Pay scales for civilians in the MoD should be the same as in the rest of the civil service. However, it is appropriate to carry out a review of civil service pay scales as a whole if the government is not able to attract the personnel it needs. Such a review is beyond the scope of this report and would require substantial separate analysis and work. Moreover, as we understand it, such a governmentwide review is ongoing.

Ukraine's MoD should seek technical assistance from NATO countries for improving personnel policy. NATO countries can continue to offer significant help in improving personnel policy, including modernizing information technology systems, reforming rank structures, and determining appropriate salary structures. This assistance is readily provided by NATO countries and could offer a concrete improvement for the MoD and GS.

Compensation System

Increasing pay, especially for service outside the ATO, will be critical for retaining personnel. Base pay for military personnel should be based on comparable compensation for civilian personnel of similar age, educational attainment, and skills. The goal of this salary structure is to attract and retain high-quality individuals by providing them with a competitive compensation package while avoiding government expenditures in excess of what is needed. Table 4.2 shows one such salary structure.

Table 4.2
Potential Compensation Structure for Ukrainian Military Personnel

Rank	Years of Service	Multiple of Lowest Pay	2015 Dollars	2015 Hryvnia
General of the Army	20	13.7	1,914	40,331
Colonel general—admiral	20	9.1	1,272	26,809
Lieutenant general—vice admiral	20	8.6	1,206	25,403
Major general—rear (counter) admiral	20	8.4	1,181	24,877
Colonel—captain 1st rank	16	5.7	799	16,826
Lieutenant colonel—captain 2nd rank	12	4.7	657	13,849
Major—captain 3rd rank	8	4.0	564	11,887
Captain—captain lieutenant	6	3.0	423	8,923
Senior lieutenant—senior lieutenant	4	3.0	415	8,742
Lieutenant—lieutenant	3	2.4	334	7,041
Junior lieutenant—junior lieutenant	0–2	1.9	266	5,596
Sergeant major—senior midshipman	8	2.6	362	7,627
Master sergeant—midshipman	6	2.2	310	6,531
First sergeant (starshina)—chief ship starshina	4	1.9	260	5,485
Senior sergeant—chief starshina	4	1.7	234	4,922
Sergeant—starshina 1st class	3	1.4	203	4,268
Junior sergeant—starshina 2nd class	2	1.3	175	3,696
Senior soldier—senior matrose	2	1.1	157	3,307
Soldier—sailor	0–2	1.0	140	2,950

SOURCE: Salary multiples were derived from U.S. Department of Defense, "Military Pay Charts: Jan 1, 2015," Defense Finance and Accounting Service, 2015.
NOTES: For each rank, we list Army ranks followed by a dash and Navy ranks. Exchange rate utilized was 21.07 hryvnia to the dollar. U.S. compensation packages for each rank and skill set are based on a comparison with civilian salaries for similar levels of experience and qualifications. For example, base pay for new recruits is based on average entry-level wages for high school graduates. The exact ratios to civilian pay shift over time based on budgetary pressures and civilian unemployment rates. Military salaries tend to fall below comparable civilian salaries when government budgets are under pressure and unemployment rates are low. They tend to rise above comparable civilian salaries during times of high unemployment or war. The U.S. 2015 salary schedule is competitive for attracting and retaining a high-quality military. Ukraine's inflation is very high, albeit now declining, and the country is in deep recession. In real terms, salaries are changing on a monthly basis because no one receives instant compensation for inflation and inflation adjustments are not taking place across the board. Thus, relative wages among occupation groups are also changing monthly. Trying to match Ukrainian civilian salaries against our proposed military salaries in this context is not very useful. Accordingly, we used the most recent figure we had available (i.e., wages for unskilled construction workers) as the minimum wage that would be needed to attract new contract soldiers. We then used multiples of U.S. salaries for the equivalent rank structure for the Ukrainian military—adjusted for average length of service, as well as rank—to provide an illustrative salary structure for the Ukrainian military. These figures will need to be adjusted for ongoing inflation.

All military personnel in the ATO, regardless of rank, should receive the same monthly combat pay. We recommend that combat pay be set at a level equivalent to monthly compensation for the lowest-paid rank in the Armed Forces of Ukraine.

Enlistment and salary supplement bonuses should be used to recruit and retain people with targeted skills. However, across all military compensation, no more than 10 percent of expenditures should be paid through bonuses.

Active-duty personnel should receive either military housing or a cost-of-living adjustment sufficient to rent housing on the civilian market wherever they are stationed, for the period when they are stationed at that location only. Housing allowances should be determined using objective, third-party measures of local housing costs. The Ukrainian military should phase out the provision of permanent housing to members of the armed forces. Current or retired members of the military who have been promised permanent housing should receive lump-sum payments or some other form of compensation in lieu of receiving apartments.

We recommend that Ukraine maintain the current military retirement system for those currently serving in the armed forces, subject to modifications forced by budgetary pressures. We further recommend that the Ukrainian government evaluate options for shifting to a compensation system for the military in which retirement forms a smaller share of lifetime compensation. The new military retirement system should be funded through joint contributions from the government and the salaries of military personnel into an independent retirement fund, and should be applied to all new personnel once it is operational.

Training

We recommend that Ukraine maintain its four training bases for initial individual and unit training, and that these be equipped with electronic training aids, such as simulators and Multiple Integrated Laser Engagement Systems. Each military occupation specialty should have its own dedicated programs at one of the four training centers.

Instructors should be combat veterans who have been trained to serve as instructors. Lessons-learned organizations should adapt lessons from combat operations into changes in future doctrine.

Procurement

What Does the System Need to Do?

A properly functioning military procurement system helps make sure that forces have what they need to fight. To do this, it needs to (1) set requirements for capabilities, quality, and quantities for arms, military equipment, and supplies; (2) allocate resources based on national defense priorities; (3) identify vendors; (4) order and take delivery of sufficient arms, equipment, and supplies to meet those requirements; and (5) ensure that items meet agreed-upon terms for quality and quantity and are purchased at the lowest prices available. The system must be responsive, transparent, competitive, and designed to inhibit corruption.

Supporting Ukrainian industry should not be a primary goal of the procurement system. In light of the security threats that Ukraine faces and its limited economic resources, procuring appropriate, high-quality arms and supplies at the lowest cost should be one of the highest priorities of the Ukrainian defense establishment. Ukraine's military industrial complex should serve that goal, not vice versa.

Problems in Ukraine's Current System

Structure

No single individual is responsible for the entire procurement and logistics process. Responsibility for requirements, procurement, and logistics is divided between the MoD and GS. Requirements for procurement are set within the GS; supplies, arms, and equipment are procured by the MoD; and the GS distributes them to the troops. No intermediary below the President has effective authority to integrate these closely related functions or to resolve disputes between the MoD and GS, as the CHoD has a separate reporting line to the President.

This division leads to two major problems: It makes it difficult to identify exactly where the system is failing and make appropriate corrections, and it prevents taking account of full life-cycle costs—including procurement, sustainment, transportation, and disposal.

Contracting

Most weapons and equipment are procured through sole-source procurement contracts through the State Defense Order. Sole-source contracts tend to result in higher prices and lower quality. Armaments and almost all equipment are procured through the State Defense Order, a classified document that is approved each year by the CoM. One of two procurement departments within the MoD—the Arms and Equipment Development and Procurement Department—carries out the classified tenders and procurement specified under the State Defense Order. This procurement is from sole sources based on a list of designated vendors, many of which are state-owned enterprises. UkrOboronProm controls many of these companies. Its influence limits the ability of the MoD to procure foreign equipment. As a result, there is little transparency in pricing, quality control, or the process through which these weapons and equipment are procured.

Proponents of the system of classified procurement argue that it is justified because the details of Ukraine's military procurement could be used by its enemies. For example, information on the number of tanks being modernized or repaired could provide Ukraine's enemies with information on the state of the Ukrainian Army. However, it is unlikely that procurement contracts would provide Ukraine's enemies with actionable intelligence. In most countries—even countries in conflict, such as Afghanistan and Iraq—information on procurements of conventional weapons is publicly available. Because the costs of lack of transparency are so high, an open-procurement system would provide greater benefits than the current use of the classified State Defense Order for most equipment and many weapons.

The process of contracting for nonlethal supplies, including food, fuel, clothing, spare parts, etc., is also flawed. The responsibility for contracting for these supplies lies with a second department within the MoD: the Public Procurement and Supplies Department. The system of public tenders in this department is opaque, unnecessarily slow, and insufficient to prevent corruption. The process of announcing, vetting, and bidding takes a minimum of three months and often longer, preventing quick responses to current needs. Vendors, often with inside information, can offer a slightly lower price at later stages in the procurement process to overturn previously announced bids. Moreover, the current requirements for documenting for vendors make it very difficult for foreign vendors to compete for Ukrainian defense contracts. As a result, Ukraine pays higher prices than necessary and could be receiving lower-quality goods or less-capable equipment than it would under a more competitive purchasing system.

Significant improvements implemented in recent months are making the procurement of nonclassified items more flexible, responsive, economical, and efficient. These improvements include

- a new system of e-procurement designed to more broadly publicize tenders and elicit more-favorable prices
- shifts toward computerized record-keeping
- reductions in the number of individuals who need to sign off on contracts, simplifying the process and making it easier to clearly assign responsibility for purchase decisions.

Systems of quality control are flawed or absent. Armaments and equipment and other supplies have been accepted even when they have failed to meet quality or other specifications of the contract, the result of an insufficiently transparent inspection and validation process. Quality assurance at time of final delivery has been conducted by military representatives of the MoD. Often, these same representatives were responsible for negotiating prices and otherwise working closely with suppliers. While this does not create an automatic conflict of interest, measures should be taken to maintain an arms-length relationship between suppliers and the MoD, as we discuss in the next section. Moreover, under the current system, it is difficult to sanction suppliers or withhold payment, even when products fail to meet contract terms for quality and timely delivery.

Individuals with conflicts of interest are in a position to affect procurement decisions. Contracts are negotiated by military representatives who tend to have long-standing relationships with state-owned suppliers, creating an environment for potential corruption. Some requirements appear to be tightly set to favor specific companies, precluding bids from competing suppliers. The system of tender committees that do not include procurement professionals facilitates corruption and inefficiency. Individuals with ties to the defense industry can guide contracts to favored suppliers without taking responsibility for the decision. Suppliers can also distort the procurement process by offering slightly lower prices through a letter to the Minister of Defense in the midst of the negotiation process with the original winner of the contract.

Standards

Continued use of Soviet-era standards hinders warfighting effectiveness, internal interoperability, and relationships with NATO suppliers and partners.

Outdated standards for equipment and materiel make it difficult to adapt to a rapidly changing battlefield. For example, the use of Western secure radio systems is technically against military regulations. While Ukraine will likely continue to use Soviet- and Russian-standard equipment for some time, some NATO-compatible standards can and should be adopted to facilitate internal and external interoperability and direct material assistance from NATO.

Recommendations

Structure

As noted, we recommend a single procurement department under an integrated MoD formed out of the current Public Procurement and Supplies Department (see Figure 4.3). The procurement department should issue tenders and carry out contracting processes for large purchases. The new department would house program offices responsible for different types of purchases, including new weapons systems, existing major weapons systems, equipment, spare parts, etc. These program offices should evaluate and undertake procurement decisions based on full life-cycle costs.

The Ukrainian government should develop a new means of organizing major purchases other than the current classified State

Figure 4.3
Proposed Chain of Command for Procurement

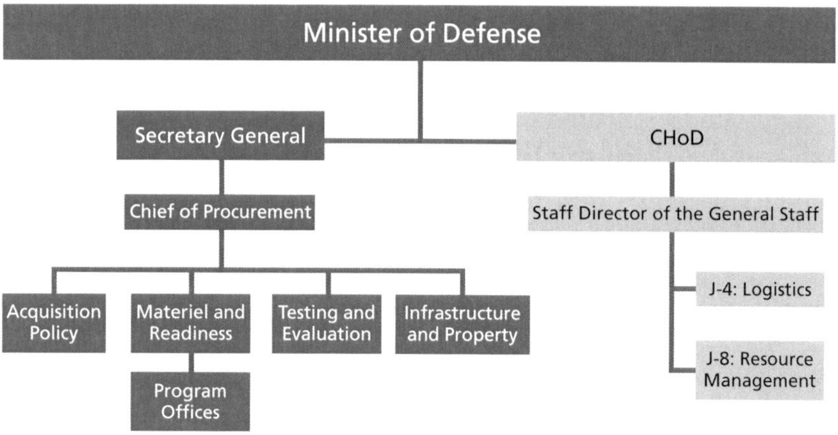

RAND *RR1475/1-4.3*

Defense Order. This system should include public notices for new requests for proposals.

Ukraine also should examine granting military organizations budgetary authority for purchasing some classes of basic supplies to ensure that current needs are met in a timely manner. This follows models in many Western countries. The J-4 Logistics and J-8 Resource Management departments could be responsible for conducting such procurement. This could also be extended to the J-4 and J-8 functions for fielded units conducting operations.

Under the new MoD structure, requirements for procurement of arms, equipment, and supplies should be developed by committees, chaired jointly by the Chief of Procurement (or designate) and the Chief of Capability Development (or designate). The committees would be composed of representatives from the Departments of Capability Development and Finance/Comptroller, the GS (including the J-4), the armed services, and the JOC. The committees could draw on civilian, and, as relevant, military contracting officers and officials from the audit department and the Finance Ministry, but will rely mainly on specialists in system engineering, technology, force structure, and operational concepts.

We recommend Ukraine adopt a quality-control system for accepting supplies, equipment, and armaments that involves physical audits of deliveries by an independent set of military officers and civil servants with technical expertise, potentially divided among various organizations and departments as appropriate for different sorts of materiel (and with committees and commissions set up in some cases). Policy and oversight for these functions would reside under the Procurement Department in a separate testing and evaluation office, which would oversee test facilities and quality-assurance procedures. Personnel might rotate between these tasks and other roles in procurement, capability, and logistics, with care taken to avoid conflicts of interest. Individuals responsible for the physical audit will have to sign off personally on the quality of the deliveries. If the deliveries are subsequently found not to have met contractual standards, the reasons for this must be investigated, potentially calling in the Office of the Inspector General. If the individual responsible failed in his or her duties, sanction by dismissal or potentially legal action should follow.

The MoD should rigorously enforce payment for final delivery contingent upon the product meeting the contracted quality and other specifications, with the caveat that for complex systems, standards must be based on performance metrics and thorough testing. Good-faith execution of quality assurance criteria must also be considered.

Contracting

All procurement should be competitively bid wherever possible, including taking bids from foreign suppliers, in keeping with Ukraine's commitments under its association agreement with the EU. Sole-source and secret procurement should be limited to those items that can only be procured from within Ukraine.

A team of individuals in the MoD should be immediately tasked with reviewing current classified and sole-source procurement to identify items and materiel that can be openly and competitively bid in a move toward transparency and competition. Once items are identified, the Public Procurement and Supply office should make those tenders available for competitive contract bids. This review should be seen as critical but not sufficient to overall procurement reform.

The Ukrainian government should expand the current e-procurement program and other projects for simplifying current tender and negotiation processes. E-procurement for commercially available items should be adopted as the standard practice.

Classifying procurement orders as secret or a higher classification should be limited to the acquisition of particularly sensitive items, such as highly advanced technologies or select special operations or intelligence items.

Careful and stringent cost analysis should be undertaken throughout the procurement system, with cost specialists responsible for vetting and reviewing procurement plans.

For those items that continue to be purchased on a sole-source basis, the Ukrainian government should ensure that it receives the same prices provided to other customers, if those are available, or prices comparable to analogous items sold by foreign suppliers. Negotiations on prices should be conducted by individuals who will not be responsible for accepting delivery of the item. Staff responsible for negotiating sole-source contracts should be rotated with sufficient frequency to maintain arms-length relationships and avoid the potential for conflicts of interest (e.g., every two years).

An oversight system should exist such that staff responsible for quality assurance are also responsible for reviewing solicitations and negotiated contracts (before execution) above a certain threshold to ensure the terms are not intended to steer work to certain vendors and result in deliveries that do not meet requirements.

Personnel within the procurement office should consist of a mix of civilian professionals and military officers. There should be strict procedures in place to avoid conflicts of interest through signed statements by civilian professionals and officers participating in the procurement process and periodic investigations of the financial affairs of these individuals.

The Ukrainian government should increase penalties for failure to meet contracted commitments in government procurement. Suppliers that have failed to meet prior commitments should be identified—and in cases of failures to meet contract specifications, precluded from future bids.

Standards and Interoperability

To ensure internal and external interoperability, and to ease cooperation and donations from NATO allies, Ukraine should adopt or add NATO standards for equipment and supplies where possible. The current pilot project to adopt new standards, led by NATO, should be expanded. This does not mean that Ukraine should jettison all existing equipment and standards, but that it should make it easier to also utilize materiel provided by or purchased from NATO and NATO-compatible countries.

Ukrainian military and civilian agencies should, where possible, seek to use compatible equipment and make purchases jointly if such purchases would result in lower prices without hampering effectiveness. This will generally make the most sense for simpler items (e.g., clothing, food). Prospects for unifying procurement of complex systems will need to be reviewed on a case-by-case basis.

Logistics

What Does the System Need to Do?

Ukraine's military logistics system needs to provide military forces with the supplies and equipment they need, when they need them. The system needs to operate efficiently and economically to minimize state expenditures. It should be designed to safeguard the state's funds and property.

Problems in Ukraine's Current System

Structure

The structure of Ukraine's logistics system is a carryover from the Soviet Union and does not follow the structure of logistics systems in NATO countries. The GS and unit structures are organized into rear and armaments staffs and units, responsible for supplies and armaments, respectively. The elements of the logistics organization were restructured over the last decade, eliminating supply units and contracting sustainment responsibilities to outside organizations. The underlying assumption under which these changes were made was that

Ukraine did not face an immediate threat, and that the logistics system did not need to supply units in combat. Since the beginning of the war, the supply units and other elements of the system have been partially rebuilt, but the old Soviet system remains, with significant limitations.

The system for determining unit requirements for supplies, equipment, and weapons is inflexible, preventing units from receiving critical materiel. Units are allocated materiel according to established requirements (or "norms") approved by the CoM. Officially, units are not permitted to receive items beyond these requirements. However, many of the established requirements are outdated or inappropriate for the current conflict. For example, at the beginning of the war, bottled water was not included in units' supplies. While some requirements have been changed, including the provision of bottled water, the policy stipulating that changes in requirements must be approved by the CoM is clearly a major impediment to a responsive logistics system. Combat needs that are not provided for under current requirements are often met by volunteers. These needs include artillery computers, secure radios, and generators. This equipment fulfills critical combat needs in the current conflict, but unit commanders must violate rules and procedures to operate this equipment and procure the relevant fuel and parts. While volunteer contributions have been invaluable to Ukraine's war effort, they are not a sustainable substitute for equipment and supplies provided through the military logistics system. Moreover, not all equipment provided by volunteers is suited for a combat environment.

Responsibility for logistics within the military is divided and unnecessarily complex. Responsibility for supervision of the logistics system is divided between the Rear and Armaments Directorates within the GS. In addition to nation-level logistics tasks, such as managing the major supply warehouses, the Rear and Armaments Directorates supervise the staffs and units that are actually responsible for transporting and managing supplies, equipment, and weaponry. Moreover, the logistics system is not well integrated with procurement in the MoD. In particular, the categorization between supplies that are transported by rear units and those transported by armaments units differs from the categorization of supplies that are purchased by the Public

Procurement and Supply Department and the Armaments Department. The complexity of the current system undermines Ukrainian and international efforts to resolve logistics problems.

The many organizations operating in the ATO have duplicate supply chains. The National Guard, State Border Guard Service, and Ministry of Interior have their own supply and logistics chains. Some of these supply chains reportedly work more efficiently than the military system, but it is not clear under what circumstances this is or is not the case. It is clear that problems of duplication arise on some occasions, while at other times the Ukrainian military reportedly provides these units with supplies when their own supply chains fail.

Effectiveness and Accountability

The effectiveness of the logistics system has improved over the past two years, but continues to be insufficient to Ukraine's needs. At the beginning of the current conflict, there was consistent reporting that Ukraine's military-logistics system was barely functional. Combat units were only supplied with food, fuel, and supplies due to the efforts of volunteers. Since 2014, Ukraine's military-logistics system does appear to function much better, as many units report that they generally receive the materiel on their lists.

However, assessments of the current ability of the system to properly supply operational units vary widely. GS headquarters personnel report that the logistics system has been responding adequately to troop needs, and that shortfalls in certain materiel are generally the result of ineffective or corrupt procurement practices. NATO and other Western advisers continue to express serious concerns over the armed forces' logistical capabilities, particularly the lack of well-managed depots close to the combat zone, shortages of supplies, and insufficient transport capacity, especially from forward supply bases to the front. Some unit commanders in the ATO continue to express serious concerns about deficiencies of logistical support to the field, in some cases noting that they had to purchase supplies from their own personal funds.

Perceptions that the logistics system fails to supply troops with basic equipment and supplies undermine public support. In some instances, erroneous information regarding gaps in supplies has contributed to diverting domestic and international aid to provide unnecessary supplies.

Ukraine currently lacks an automated, networked inventory control system for the armed forces that includes a complete inventory of supplies and tracks movements of supplies in and out of the system. The current paper-based process is cumbersome, slowing procurement and precluding delivery of supplies in a properly sequenced, timely fashion. Transactions and inventories are formally managed on paper throughout the system, and computers are used within individual units or departments to track the flow of paper orders and inventories, but they are not linked to a central system. In the absence of a centralized automated system, logistics personnel waste time, money, and effort struggling to anticipate the needs of units, locating requested items within the system of warehouses and supply depots, and planning and executing the provision of supplies across the security forces.

The existing system also contributes to a lack of transparency concerning the cost, sources, and ultimate disposition of materiel. A reliable automated system is needed to establish a clear chain of custody for equipment and armaments. This is also necessary to provide the accountability that Western donors require for their assistance.

Recommendations

Structure

We recommend that a single Logistics Department (J-4) be created in the reformed GS under the Staff Director of the GS. The J-4 would have responsibility for managing national-level logistics functions currently in the GS. It would run the warehouses, associated transport units, and the future inventory management system, and would work closely with the Department of Procurement under the Secretary General to ensure that immediate supply needs are met. In coordination with the J-4, the logistics staff and units under the JOC and the armed services would be responsible for managing the day-to-day sustainment of military units.

To ensure that the J-4 and the Departments of Procurement and Capability Development work closely in the formulation of requirements for supplies and equipment, joint committees or boards need to be created, with representatives from each organization. Lines of communication need to be set up to ensure that units responsible for combat logistics are linked up with those responsible for defining the characteristics and quality of purchases needed to meet the needs of combat units.

Creating a more efficient and flexible system for changing supply and equipment norms is critical, given the limited resources and poor prewar condition of the Ukrainian forces. Responsibility for setting norms would reside with the CHoD (delegated to the GS J-4) for noncombat units and with the Commander of Joint Operations (JOC J-4) for units in combat. Units should be given explicit encouragement to improvise and take advantage of useful equipment that is not officially listed in the norms. The MoD and GS should examine the feasibility of providing each unit with a small budget to meet urgent needs.

The military logistics system, the largest logistics system for Ukraine's security forces, should have the capacity and authority to provide supplies to all government agencies involved in a joint operation with military forces. At a minimum, its supply chains should be closely integrated with the supply chains of other government efforts in an area of operations. Local logistics commanders should have wide authority to organize and distribute supplies across different government agencies.

Effectiveness and Accountability

Ukraine should assess the current effectiveness of the logistics system and carry out regular reviews. While recognizing gaps may be temporarily embarrassing, public and regular assessments of how the current system is functioning are critical for making improvements and gaining support from the public and international community.

As soon as possible, Ukraine needs to purchase and install a single inventory control system that tracks inventories; requests for supplies; incoming purchases; and deliveries for all supplies, equipment, weapons, and other materiel. This effort should expand on the ongoing NATO pilot project. Logistics personnel should be able to obtain information from throughout the system on current stores, incoming orders, requests, and deliveries. Such a system would help prevent corruption and loss and would ensure that Ukraine can account for donated military supplies, equipment, and weapons.

Ukraine should review the level of classification of information on supplies. It is much easier and more efficient to operate a computerized logistics system in an open environment than through a classified system.

The Armed Forces of Ukraine should provide more-transparent information to the public about the provision of supplies to combat forces so as to counter the flow of partial and inaccurate information from various sources about Ukrainian government support for its troops.

The MoD should work with foreign partners to provide training programs for field logisticians appropriate to the challenges faced in modern combat environments. Military supply officers should be provided with opportunities to participate in events and attend partner nations' logistical centers of excellence to gain exposure to a variety of professional best practices.

Ukraine should consider adoption of electronic identification cards and electronic readers to facilitate greater accountability for the use of supplies and equipment. For instance, provision of electronic identification cards to individual soldiers would allow for more-precise accounting of the distribution of individual equipment and supplies in a barracks environment. Use of electronic signatures in the transfer of equipment and supplies at different points in the logistics system would create greater transparency over the location and possession of materiel.

CHAPTER FIVE

Cybersecurity

In using the term *cybersecurity*, we adopt the definition commonly used in the broader Euro-Atlantic community: the preservation of the confidentiality, availability, and integrity of information in cyberspace. Our focus is on defensive cyber capabilities (i.e., prevention, detection, responding, recovering, and learning from incidents and breaches) and not on the development of offensive cyber capabilities.[1] As of the fall of 2015, those organizations involved in cybersecurity in Ukraine included the Ministry of Justice, the MIA, the Ministry of Economic Development and Trade, the SBU and Ukraine's other intelligence agencies, the SSSCIP, and the Ministry of Defense.

National Cybersecurity Strategy and Concept

What Should the System Do?

A national cybersecurity strategy is a fundamental document required to shape a country's national approach toward its cyber ecosystem, as well as to increase its preparedness level to face threats to, as well as serious risks and breaches in, its networks and information systems. The overarching aim of a cybersecurity strategy is to set strategic goals

[1] Different approaches have been taken to facilitate conceptualizing and understanding the cyber domain. As background for our mapping and gap analysis of Ukraine's national cyber ecosystem, we employed the University of Oxford's Global Cyber Security Capacity Centre Maturity Model to identify the different issue areas and capability requirements characterizing the cyber ecosystem. See Global Cyber Security Capacity Centre, "Cyber Security Capability Maturity Model (CMM)—V1.2," University of Oxford, 2014.

so that decisions and developments in the national cyber domain are not shaped by narrow bottom-up processes. A strategy should be based on a whole-of-government approach with the goals of protecting critical national infrastructure, developing incident response capabilities to mitigate the impact of breaches and malicious attacks, creating a legal framework that tackles cybercrime while establishing an environment conducive to opportunities offered by information communication technologies (ICTs) for economic growth and development, and participating with international peers to the advancement of the cyber field.

Problems in Ukraine's Current System

For the past three years, the Ukrainian authorities have been developing a national cybersecurity strategy. Consultations around the content and aims of the strategy have occurred through informal mechanisms or on an ad hoc basis, rather than through a systematic, coordinated process bringing together all parts of the government. The process of writing a national strategy has dragged on as multiple drafts have circulated among stakeholders who do not have a clear understanding of timelines or of who is in charge of developing the document.

Recommendations

We recommend that Ukraine define a cybersecurity strategy, drawing on the existing framework for the development of National Cyber Security Strategies of the European Union Agency for Network and Information Security.[2] The strategy also needs to emphasize the protection of human and civil rights in cyberspace.

[2] European Union Agency for Network and Information Security, *An Evaluation Framework for National Cyber Security Strategies*, Heraklyon, 2014.

Cybersecurity and Defense Organizational Structure

What Should the System Do?

There is currently no single organizational structure for cybersecurity that is considered best. However, a number of underlying shared principles exist and should be taken into consideration when devising a structure for cyber defense. One is to establish a whole-of-government approach, in which public sector actors can work in an environment that reconciles different goals and perspectives; harmonizes tasking across the strategic, operational, and tactical levels; and avoids duplication of effort and stovepiping. Such an approach would also facilitate cooperation with organizations outside the government.

Problems in Ukraine's Current System

The organizational structure and coordination of tasks and responsibilities among Ukrainian organizations in cybersecurity and cyber defense have not yet been fully developed. Overall, there are no organizational principles or governance mechanisms in place to facilitate a whole-of-government approach. Stovepiping appears to be a significant problem, as a multiplicity of organizations and bodies often work in parallel on overlapping, if not identical, areas. Sometimes this appears to be the product of a lack of mutual awareness and communication, but in certain instances, duplication of effort seems to be the result of turf wars and competition for responsibilities over strategic cybersecurity and cyber defense areas. For example, the SBU and the MIA have nearly identical responsibilities for forensics related to investigation of cybercrimes, and no criteria could be ascertained with regard to division of work and tasks between the two institutions.

Due to constraints on salaries, only a limited number of skilled individuals working on cybersecurity and cyber defense matters appear to be employed within public sector institutions. Moreover, these capable individuals are scattered across a number of organizations, preventing any single Ukrainian institution or body operating in the cyber domain from achieving a critical mass of maturity and capabilities from either a human or technical point of view. In fact, even when the staff is capable, they have limited or no access to state-of-the-art kits and tools.

Recommendations

We recommend the creation of an interagency cyber coordination committee, the Joint Committee on Cyber Security (JCOCS) under the auspices of the NSDC, modeled on the JCOI. The JCOCS would be responsible for coordinating Ukraine's cybersecurity activities, including the development and continuous review of a national cybersecurity strategy and concept. It would need to engage in defining the precise role of the different government agencies involved in cybersecurity to deconflict mandates and responsibilities and help avoid unnecessary overlap and duplication of efforts.

The rationale for establishing a new body to coordinate cyber activities in Ukraine, rather than assigning this function to an existing one, is twofold. First, this would ensure that an independent organization is tasked with coordination activities, thereby reducing the likelihood of conflicts of interest. Second, placing the JCOCS under the NSDC would ensure that cyber coordination remains an independent and continuing function, regardless of any restructuring in other parts of the Ukrainian security sector organizational landscape.

We also recommend that the responsibilities of the different ministries and agencies involved in cybersecurity be clarified. The Ministry of Justice would take overall responsibility for creating a cyber legal framework and structure, with the General Prosecutor of Ukraine overseeing its enforcement in cooperation with law enforcement agencies. The role of the MIA would be limited to that of coordinating cyber incident management with more-general crisis management measures in the face of large-scale crises, be they natural or the result of human action. The Ministry of Economic Development and Trade would assist in the development of the cyber legal framework that addresses matters such as intellectual property rights, and would play a role in facilitating cybersecurity awareness-raising activities and training, specifically in relation to the private sector. The Ministry should also formulate cyber technology standards fit for security purposes in concert with industry actors who manufacture the relevant technologies.

Critical National Infrastructure Protection

What Should the System Do?

Shared responsibility between governments and the private sector for the achievement of cybersecurity and resilience is deeply entrenched. Over the years, owners of strategic infrastructure assets and providers of fundamental services have seen their roles and responsibilities progressively increased as they became more embedded in national security alongside government actors.

Work on critical infrastructure protection in the cyber domain mostly occurs in the prevention phase of cybersecurity and defense activities. As a first step, a review and identification of what constitutes critical national infrastructure within a country and what dependencies this infrastructure has on ICTs is necessary. Second, national stakeholders need to identify the type of threats to which national critical infrastructure is exposed, identifying the different types of actors from which threats may arise, their ultimate goals, and the means they can employ to achieve them. Critical infrastructure can be subjected to a variety of attacks with aims ranging from noise generation to information disruption and information subtraction. A variety of groups might also be targeting critical infrastructure, from individuals and "hacktivist" groups (cyber activism), to cyber criminals (cybercrime), to state-sponsored attackers (cyber terrorism or cyber espionage). A fuller understanding and characterization of the threat landscape is the first step toward adoption of suitable strategies to mitigate risks and threats to critical national infrastructure.

Strategies should incorporate state-of-the-art network protection tools and techniques, as well as mechanisms to ensure continuous reassessment. Based on the type and level of sophistication of threats, additional measures and strategies may be required, such as the development of appropriate counterintelligence activities, based not only on cyber information-gathering techniques, but also on other approaches, such as traditional human intelligence and signals intelligence undertakings.

To achieve protection and promote involvement from private sector actors, governments have taken significantly different routes to ensure that adequate protection is in place for critical national infrastructure. A review of existing rules and regulations shows that they range across a spectrum of approaches characterized by different degrees of enforceability. Existing government regulations vary, from encouraging self-regulation of the private sector (due to the highly dynamic and technical nature of requirements); to adopting soft (i.e., nonbinding) regulations aimed at encouraging good practice; to adopting laws prescribing clear minimum standards for technology deployment, internal security controls, requirements for notification of security breaches and disaster recovery, and business continuity plans.

Problems in Ukraine's Current System

At the national level, the Ukrainian government is not systematically managing the security of national critical infrastructure. No agency seems to be tasked with overseeing and managing this function, maintaining situational awareness, or devising protocols for securing networks. The SSSCIP, whose mandate touches most closely on critical infrastructure protection, appears to be maintaining a narrow approach in its work. In fact, the SSSCIP focuses mostly on special communications' technical matters (e.g., establishment of secure communication lines with foreign top officials, provision of mobile service for state and governmental authorities, and cryptographic information protection), rather than on the broader spectrum of threats and concerns characterizing critical infrastructure protection at the cyber level.

We found neither a clear definition of what constitutes national critical infrastructure nor a clear cyber system for its protection. The Ukrainian government does not appear to have the legal authority to enforce cybersecurity standards and requirements on the private sector prior to an incident, although incidents can quickly escalate to a national crisis, at which point the government would have to step in. However, the Ukrainian government does have legal authority within the telecommunications industry to enforce regulations on cybersecurity under the Law on Protection of Information, passed in 2006. The Law on Telecommunications also imposes obligations concerning cybersecurity on the industry.

Even within the public sector, no protocols are in place to secure networks and protect data, although the SSSCIP has produced guidelines indicating priority areas for action and measures to be undertaken. Budgetary constraints are likely to continue to encourage organizations to employ old kits and software of dubious origin.

Recommendations

Ukraine needs to give greater attention to protecting its critical national infrastructure by first identifying what assets constitute that infrastructure based on clearly defined criteria. Identified assets then need to be assessed for their critical dependencies on ICTs and cyber-related technologies and their level of criticality within the national systems. Such a list should be subjected to continuous review, updating, and reassessment to retain strategic relevance and importance.

Once a clear list of critical national assets needing cyber-based protection measures has been compiled, different agencies need to be tasked with aspects of critical infrastructure protection according to different phases of the management cycle. In particular, the security services need to be tasked with carrying out risk and threat assessments for infrastructure assets, as well as gathering information and intelligence, engaging in cyber espionage, and collecting cyber-attack intelligence. JCOCS needs to set up mechanisms to exchange information within the government to avoid lack of trust among different security services. The Ministry of Economic Development and Trade should work to establish public-private partnerships and help develop a legislative framework to encourage the protection of critical national infrastructure from the owners' side (e.g., establishment of clear minimum standards for technology deployment, internal security controls, requirements to notify the government of security breaches, disaster recovery, and business continuity plans).

Incident Response

What Should the System Do?

Incident response is normally the responsibility of a technical agency operating as a national or governmental Computer Emergency Response Team (CERT). A national or governmental CERT needs to act as a service provider for a variety of stakeholders and constituencies. Among its tasks, a CERT would provide a reliable 24-hour point of contact for emergencies and incidents; facilitate communication among experts involved with incident response; act as a central hub for researching, gathering, and distributing information on vulnerabilities, breaches, and threats among constituencies and stakeholders that the CERT serves; and promote the establishment of other CERTs within a variety of constituencies and stakeholder groups or organizations so as to establish a national network of CERTs and strengthen incident-response capabilities.

Problems in Ukraine's Current System

The Ukrainian cyber landscape is characterized by insufficient incident response capabilities. CERT-Ukraine (CERT-UA) is currently nested within the SSSCIP, limiting its role and capabilities to that of a subagency. Although stakeholders we interviewed seemed to have a clear understanding of the current cyber threats, information on the role played by CERT-UA in maintaining such awareness and providing information on the technical tools to respond appears to be nonexistent. However, a proposal by the SSSCIP to put forward CERT-UA as a possible recipient of funding and materiel from the NATO Cooperation funds represents a step in the right direction for establishing mature incident response capabilities inside Ukraine.

CERT-UA notwithstanding, Ukraine appears not to have CERTs that serve the private sector, other bodies, and constituencies of the public sector. This results in a lack of mechanisms through which to share information on cyber vulnerabilities and threats among Ukrainian actors and stakeholders that depend on the cyber domain.

CERT-UA is also not currently liaising with security services to obtain up-to-date, reliable intelligence on critical infrastructure assets, possible threats menacing them, and approaches to mitigation to be employed to tackle threats, raising doubts as to the situational awareness of Ukrainian public sector bodies and institutions in the cyber domain. Therefore, there appears to be limited cyber support for the private sector and other bodies and constituencies of the public sector.

Recommendations

We recommend that the CERT-UA be moved out from under the SSSCIP so that it is a fully autonomous organization that would be responsible for all civilian aspects of cyber incident management and response and able to support and liaise with all government agencies, the JCOCS, and other organizations in incident response operations.

The rationale for detaching CERT-UA from the SSSCIP is twofold. First, this would ensure the establishment and development of an organization, disengaged from ongoing and potential turf wars and focused only on the protection of the national cyber environment. Second, removing CERT-UA from the SSSCIP would allow for broader analysis beyond the narrow scope of special communications' technical matters in favor of taking into account the whole spectrum of threats, vulnerabilities, and potential exploits targeting users and stakeholders of the national cyber ecosystem. Resources would need to be allocated to ensure that CERT-UA could undertake these functions.

CERT-UA would need to develop technical and human capabilities adequate to act as a full-time service provider, serving all Ukrainian institutions involved with cybersecurity and offering real-time advice on how to respond to ongoing incidents, vulnerabilities, and threats. CERT-UA will need to promote the creation of CERTs within public and private sector organizations, and to facilitate collaboration and the exchange of information. There is an especially great need for skilled staff and adequate retention policies for highly qualified personnel. CERT-UA also needs to provide specialized training courses on incident management.

The JCOI would coordinate data and intelligence-gathering activities, collating and harmonizing inputs from the Foreign Intelligence Service, the Military Intelligence Service, the State Border Guard Intelligence Service, and the SBU. The Foreign Intelligence

Service, in particular, would be tasked with gathering intelligence on cyber espionage and cyberattacks from abroad. Additional data and intelligence on national critical communications infrastructure would be provided by the SSSCIP for special-communication assets and by the SBU for all other critical national infrastructure assets that have a cyber-dependency.

CERT-UA would then receive all information and intelligence on pending threats to critical infrastructure gathered and processed by JCOI to support its incident prevention and handling mission. To facilitate indirect information-sharing and trust-building between different security services and CERT-UA, JCOCS would also become directly involved, under the umbrella of the NSDC, liaising with JCOI to help establish procedures and mechanisms to regulate the flow of information to CERT-UA.

The MoD would continue to run its own military CERT to respond to incidents. However, the capabilities of the military CERT would be expanded and work closely with the CERT-UA, including information-sharing on threats and vulnerabilities, protection measures to be adopted during ongoing incidents, etc. The military intelligence service would be responsible for identifying infrastructure assets that are critical to military operations and need to be protected from cyber threats.

Crisis management would be the responsibility of the MIA, working with CERT-UA when it comes to cyber-related aspects of crisis management operations. Only in exceptional circumstances would the defense establishment be involved in crisis management.

Military Cyber Defense and Cybersecurity

What Should the System Do?

Military cyber defense refers to the capability of the military to protect its own networks and ICT-based systems (e.g., weapons systems and communication). Within this capability area, the military should be able to manage incident response; critical infrastructure protection; training and personnel; and research and development, equipment, and materiel. The military needs to employ such capabilities only in its sphere of competence (e.g., training for military personnel, rather than for the broader population) and over those critical infrastructure assets and networks that are crucial to its mission. In addition, the military may be called upon to develop and deploy battlefield-specific cyber capabilities at the tactical and operational levels and to provide ad hoc support to civilian authorities during cyber crisis management operations, including man-made disasters. Although members of the Euro-Atlantic community traditionally consider crisis management a civilian responsibility, a number of countries do prescribe a limited role to the military in a number of situations.

Problems in Ukraine's Current System

The absence of an overarching coordination body responsible for cyber defense matters within the military and the MoD hampers the development and deployment of capabilities across the Ukrainian military cyber defense system. The proliferation of departments and units trying to develop competing cyber capabilities and responsibilities contributes to this problem. In this regard, the MoD and the military are plagued by many of the same problems as the civilian cyber domain. Dysfunction, stovepiping, detrimental competition, and turf wars hamper effectiveness within the military cyber sphere just as they do in the civilian sphere.

At a doctrinal level, multiple stakeholders in the cybersecurity realm appear interested in developing tools to monitor content, which is considered outside the remit of cybersecurity and defense by members of the Euro-Atlantic community.

The Ukrainian military has reportedly set up a military CERT to respond to incidents, but it is unclear whether it is currently able to meet the needs of the military. A plan to establish a new cyber rapid response team that would respond to all the needs of the MoD and the Ukrainian military is currently being discussed.

Several stakeholders expressed a high level of trust in the air-gapped networks employed by the military. These networks would make malicious attacks from a range of adversaries more difficult because of the separation of the network from other networks (e.g., the

Internet). However, air-gapped networks have proven to be vulnerable to threats that take advantage of physical vulnerabilities in the network, poor levels of cybersecurity awareness, or practice of users (e.g., individuals who use USB drives on personal devices and then on terminals connected to the air-gapped network).

Cyber capabilities appear to be very thin in Ukraine at the tactical level of operations, limiting the ability of the Ukrainian military to take advantage of cyber technologies to bolster kinetic operations. A number of volunteers appear to have been invited to become embedded in battalions and have been providing free cyber equipment. From a cybersecurity perspective, such activities raise concerns. The volunteers could pose an insider threat, as materiel provided has not gone through a standardized acquisition procedure or review of its known and potential vulnerabilities.

Recommendations

We recommend that Ukraine create a cyber command that would conduct full-spectrum cyberspace operations, with a focus on the most severe threats, including cyber warfare, the corruption of defense networks, and the disabling of military coordination systems. The command would also protect critical infrastructure, including logistics and cyber systems supporting C3I.

A cyber command would be created within the MoD out of existing agencies within the defense intelligence community, and should be collocated with relevant intelligence units. A cyber command would have to provide overall compensation competitive with the private sector. Figure 5.1 provides a schematic overview of the resulting national cybersecurity and defense organizational structure based on recommendations provided in this chapter.

Figure 5.1
National Cybersecurity and Defense Organizational Structure

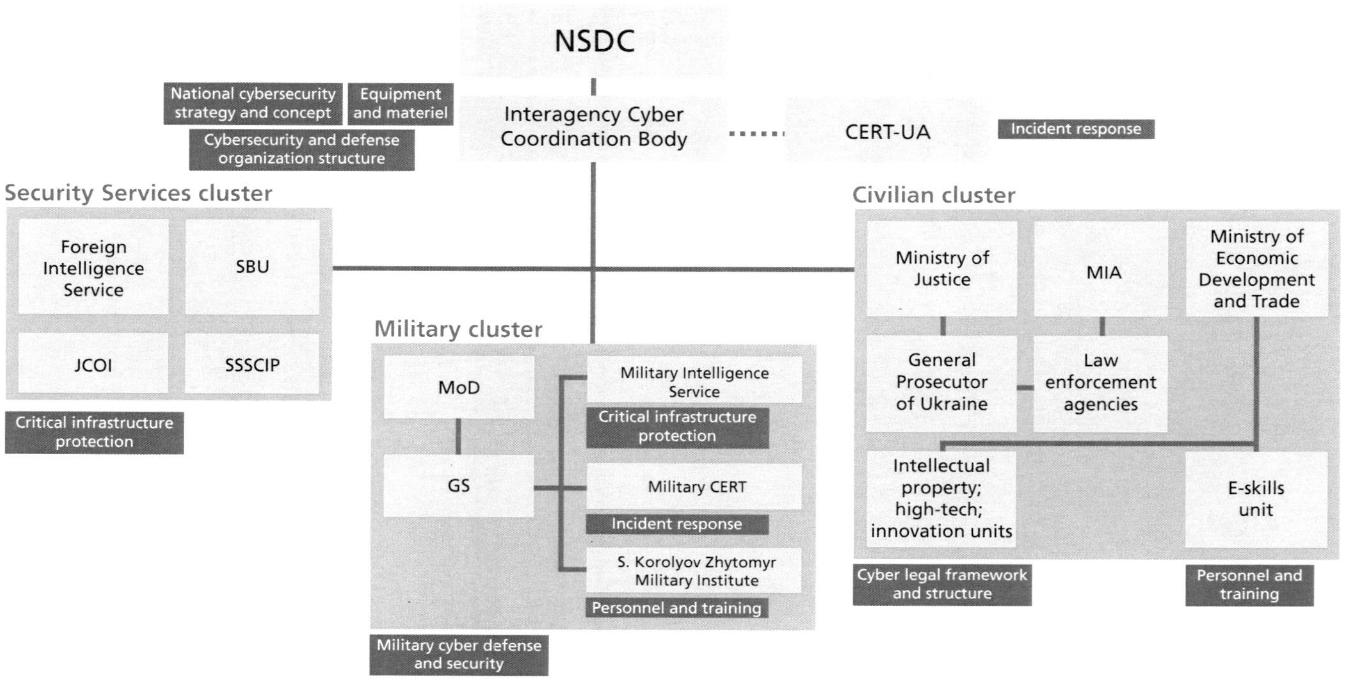

RAND *RR1475/1-5.1*

CHAPTER SIX

Defense-Technical Cooperation with Global Partners

There are significant barriers to defense-technical cooperation—which we define as the transfer of defense equipment, weapons or technology and associated information, and support—between Ukraine and foreign partners. Partly as a result of these barriers, transfer of critical weapons, equipment, and technology from Western partners since April 2014 has been relatively limited. Since shifting away from exports to Russia in 2014, Ukraine's defense industry has had difficulty finding new partners, securing greater investment, or establishing joint ventures with Western firms. Improving defense-technical cooperation could be helpful with filling gaps in weapons and equipment needed for the current conflict, and with developing Ukraine's defense industry. This chapter identifies actions that Ukraine can take to improve defense-technical cooperation.

There are several types of relevant defense transfers.

- **Assistance** refers to the transfer of defense items without any cost to the receiving country. These may be donations of old or excess defense items.
- **Sales** are purchases of defense items or technology. Sales may be directly between governments, firms, or some combination of the two. In some cases, partial grants or discounted government-supported loans may be available.
- **Joint ventures** are international commercial cooperation between companies. This may include foreign investment; coproduction; or the transfer of equipment, technology, or intangible intellectual property to further a commercial enterprise.

To date, Ukraine has received the greatest defense transfers in the form of assistance. There have been some limited sales, especially of items that used to be imported from Russia, and some preliminary discussions about joint ventures.

Many source countries of defense items and technologies share similar principles that guide their decisionmaking on whether to approve transfers, including strategic and political interests as well as the presence of safeguards necessary to prevent retransfer and misuse. Given the difficulty of affecting the broad political and strategic calculations of source countries, improving safeguards is probably the most direct route for Ukraine to improve the prospects for cooperation. Improving safeguards involves working closely with regulatory officials in the United States and other source countries, and paying close attention to the gaps and problems in Ukraine's processes, procedures, and institutions discussed later in this chapter.

Addressing Source Country Concerns

In pursuing defense-technical cooperation, Ukraine must bear in mind that source countries carefully control the transfer of defense items and technology. They make decisions regarding approval of transfers based on whether they believe it is in their national interest to do so. To this end, countries first consider the overall political and strategic implications of a transfer. Transfers could be approved to aid an ally in need, or they could be rejected over concerns that a transfer would undermine the source country's security or the security of its allies, or over human rights concerns. Source countries also consider the safeguards in place in recipient countries aimed at preventing the retransfer or misuse of defense items. In the absence of robust re-export controls, defense items could wind up in the hands of potential adversaries or other unintended recipients. High-tech defense articles may also be studied and reverse-engineered, thereby squandering the considerable investment the source country made to develop those technologies and potentially undermining the security of the source country itself.

Political and strategic concerns typically dominate decisionmaking. Even countries that place a high value on limiting the risk of misuse of defense articles and technologies, such as the United States, can decide to approve transfers in the absence of credible safeguards, given sufficiently urgent strategic concerns, although recipient countries must be expected to abide by international agreements and domestic laws that place clear limits on exports. However, in cases where political and strategic considerations are less acute, the presence of safeguards can be critically important. Stronger safeguards help to assure the source country that the political, strategic, and economic risks of approving the transfer are manageable, thereby speeding and easing the approval process. The importance of safeguards grows in the case of the transfer of advanced or sensitive defense articles or technology.

The precise safeguards that Ukraine or other recipient countries must put in place to satisfy exporter concerns vary depending on the exporter, the recipient country, and the nature of the defense item or technology. On one end of the spectrum, advanced military technologies, such as certain types of missiles, generally require very rigorous controls and assurances. On the other end, dual-use goods, such as computer equipment, typically still require a license and assurances regarding the end-user of the equipment or technology but raise fewer overall concerns. Across all types of defense items or technologies that are being considered for transfer, recipient countries such as Ukraine are evaluated based on three broad, interrelated concerns:

1. **Re-export controls.** Source countries are concerned with the potential for the items or technologies to be transferred to other states and out of the hands of potential adversaries. Effective export licensing and border control institutions are important for assuaging these concerns.
2. **End-use controls.** A license or other agreement to transfer a defense item strictly defines who may use the item and for what purpose. These restrictions reflect political concerns that the items or technologies could be used by militias or other nongovernment actors, that they could be used against civilian populations or contrary to human rights standards held by the source

country, or that the technology within an item may be studied or reverse-engineered, reducing the military or commercial value of the item.

3. **Security and bureaucratic controls.** To help ensure adherence to the previous two principles, exporting states are concerned about the controls and protections placed upon defense items and technologies after their transfer. These protections may include a variety of bureaucratic and technical systems, including inventory controls, physical security, disposal techniques, and training for government employees or military officers on the necessary security requirements and systems.

There is no universal checklist of procedures and reforms that Ukraine or any other potential recipient country can implement to ensure that the concerns of exporting countries are satisfied. Instead, Ukraine will likely need to demonstrate its concern for and implementation of these principles. Different types of safeguards may also be required for defense items or technologies of different levels of sophistication or sensitivity.

Impediments to Defense-Technical Cooperation in Ukraine

Ukraine faces two sets of impediments to defense-technical cooperation. The first set relates to the transfer of items through sales and assistance. These problems generally do not block all transfers, although they do reduce trust and make some forms of assistance and sales difficult. The second set relates to joint ventures, including foreign investment and technology transfers. Addressing these problems will likely necessitate a major overhaul of the relevant government agencies and the defense industry.

Transfers of Defense Items Through Sales and Assistance

Many countries are actively supporting Ukraine with weapons and other military equipment, including the United States, Canada, Lithuania, and the United Kingdom. By and large, Ukraine's processes, legal structures, and institutions do not prevent these countries from transferring most defense items, with the exception of some high-tech weapons and equipment. Some Ukrainian institutions, such as the export licensing office, have been substantially reformed through close consultation with Western donors and function well. The main limitation on receiving additional defense items at the present time appears to be a policy decision within Western countries not to provide lethal items to Ukraine.

However, problems and gaps in Ukraine's current systems make it more difficult, frustrating, and expensive for Western countries to provide assistance and approve the sale of high-tech weapons or equipment to Ukraine, and for Ukraine to purchase defense items from abroad. It is also likely that negative perceptions of how Ukraine has managed defense assistance to date influences how foreign companies consider investment or cooperation with Ukraine's defense industry.

Internal Coordination

The greatest problem and frustration for countries considering providing Ukraine with defense assistance is internal coordination within the Ukrainian government. Countries can easily become frustrated or deterred from future transfers by the widespread perception of dysfunction within the Ukrainian government.

- **Requests for assistance.** Many different organizations—including departments within the MoD and GS, the National Guard, and so on—make requests of foreign governments for assistance. Multiple and conflicting requests make it difficult for potential donors to evaluate where assistance could best be used. The duplication of departments for international cooperation within the MoD and GS compounds this problem. In many cases, requests are made for advanced equipment without taking into account the challenges of how an item would be used, maintained, or allocated. For example, requests for unmanned aerial vehicles are not supported by an analysis of the capabilities necessary to analyze and disseminate newly gathered intelligence.

- **Approving and signing agreements for cooperation.** Foreign officials from several governments complained that once cooperation between Ukraine and their country has been agreed upon, such as joint training or some form of assistance, it is difficult to finalize a formal agreement. Ukrainian officials at times offer different opinions concerning what level of official is permitted to sign an agreement. Uncertainty about whether delegation or approval is required by the Prime Minister, CoM, or other senior officials can slow international assistance and deter Western countries from providing additional assistance in the future.
- **Customs clearance.** Officials from several Western countries were frustrated by delays in clearing materiel through customs. Some blame the requirement that customs duties be paid even for foreign military assistance in cases where the Ukrainian government is responsible for paying some of the costs of transporting or maintaining the equipment. Others note that Ukrainian customs is not included in discussions about foreign assistance, and so its officials are unaware that donated equipment is arriving. Hence, some delays could likely be alleviated by better interagency coordination.
- **Expressions of appreciation for foreign assistance.** Representatives of some countries supporting Ukraine said that their assistance sometimes does not appear to be appropriately appreciated. This perception appears to stem from disorganization in responding to Western offers of assistance, including last-minute requests by Ukrainian officials for changes in plans, difficulty scheduling meetings, and a general sense of disarray. The perception that assistance is not appreciated could filter back through Western governments and make them less likely to provide additional assistance.

End-Use Monitoring and Perceptions of Loss or Corruption

There is a strong perception among Western donors that assistance is not well used, with concerns about items being stolen outright or diverted for personal use by senior officials and placed into storage rather than used at the front. While perceptions of misuse, diversion, and theft are troubling for countries offering assistance, the risk that materiel is not ending up in the hands of its intended recipients is especially a problem in cases of high-tech items. Given that the United States, in particular, has strict regulations and procedures to ensure that its technology remains secure and in the possession of the intended recipient, this deficiency could result in the United States limiting its assistance.

Ukrainian officials insist that they have kept records to account for all items given as assistance. They also note that many accusations of misappropriation are Russian propaganda, and that, in some cases, the loss of equipment was unavoidable due to combat conditions. There is good reason to believe that some of the worst accusations of Ukrainian misuse or mishandling of donated equipment are indeed exaggerations or outright fabrications.

However, as discussed in Chapter Four, Ukraine's paper systems for tracking equipment are outdated and vulnerable to corruption. The records from Ukraine's current logistics and tracking systems would not be persuasive to Western officials, even if they had easy access to them. Furthermore, it appears that the additional checks required under contract with Western donors for certain high-tech items, such as night-vision devices, are not being conducted. The perception of misuse or corruption, whatever the reality, is sufficient to deter donors that might otherwise provide free equipment or supplies, and to make U.S. or other officials concerned that Ukraine cannot be trusted with high-tech systems.

Flawed Import Systems

In theory, it is legally possible for the MoD and other agencies to be issued a license to import items, but in practice, UkrOboronProm has the sole ability to import defense items for use by the Ukrainian military.[1] This is problematic for at least two reasons.

1. Some potential suppliers have legal frameworks that make it difficult to contract with UkrOboronProm. In the case of the United States, for example, foreign military sales can only be concluded with a procurement authority under the MoD of the

---

[1] In July 2014, helmets and body armor were exempted from import duties and rules governing international military transfers. See "Parliament Simplified Import of Medicines and Bulletproof Jackets to Ukraine," *Ukr.Media*, July 1, 2014.

receiving country, not a state-owned enterprise. While direct commercial sales of military items from U.S. defense companies are possible, U.S. firms in the short term would likely only consider selling weapons or military equipment to Ukraine through foreign military sales given the significant political and economic risks, as well as concerns about fulfilling strict U.S. regulations about the transfer of high-tech equipment abroad. Hence, under current circumstances, most transfers of U.S. equipment and weapons are limited to assistance, rather than sales.

2. UkrOboronProm has a conflict of interest. Its subsidiaries manufacture equipment for the Ukrainian military. Consequently, it has a disincentive to import items that might be supplied by its subsidiaries. UkrOboronProm has a reputation for excessively marking up import costs by 5–20 percent or more. By making it more difficult to import items, UkrOboronProm may hope to develop Ukraine's own defense industry. While this may be an understandable, if not strategically optimal, prioritization during peacetime, during wartime it prevents Ukraine from acquiring needed equipment.

Transfer of Defense Items Through Joint Ventures

Some Western arms companies are interested in cooperating with Ukraine, including purchasing Ukrainian items or technologies, selling to the Ukrainian government, and setting up joint ventures with Ukraine's defense industry. However, foreign defense companies seeking to invest or do business with Ukraine face major challenges. Based on discussions with Western officials, defense industry representatives, and analysts, major changes will be necessary if Ukraine hopes to attract substantial investment from U.S., European, or other foreign companies. The central problem is Ukraine's overall business climate, including its regulatory regime and business culture. Political risk is another factor. Problems emanating from overregulation and partial enforcement appear particularly challenging for foreign firms seeking to do business in Ukraine. Ukraine is making changes to improve its business climate as part of its larger reform agenda, but existing challenges also affect the calculations of potential partners of the defense industry.

The development of close cooperation of the defense industries of former Warsaw Pact members with those of the United States and other NATO members offers some perspective on the potential for future cooperation with Ukraine. Poland went through a five- to ten-year process of revising its legislation and changing its institutions to align with EU standards and meet U.S. regulatory concerns about export controls, end-use monitoring, and technology security. Ukraine has already taken significant steps, including accession to multilateral arms control regimes and development of its export control system. However, Ukraine faces a more difficult political environment for defense cooperation with the United States and other NATO countries than Poland, the Czech Republic, and other Warsaw Pact members because those countries were on a clear path to NATO membership.

UkrOboronProm

The legal framework, organizational structure, and culture of UkrOboronProm are frequently cited by potential partners, foreign officials, and others as major constraints on and impediments to future cooperation.

- **Legal framework.** There is a wide consensus, including within UkrOboronProm, that the current legal framework discourages potential investors. For example, existing law requires that the Ukrainian state retain at least a 50-percent share in any venture involving state property, and stipulates that CoM approval may be necessary for ventures involving strategically important property.
- **Transparency.** For foreign defense companies to be interested in cooperating with a Ukrainian company within UkrOboronProm, the role of the consortium needs to become more transparent. In particular, foreign companies need to understand the legal framework under which UkrOboronProm operates, the organizational structure of UkrOboronProm and its subsidiaries and affiliates, and how UkrOboronProm functions as the parent company. Currently, the size, complexity, secrecy, and bureaucracy of

UkrOboronProm make cooperation with the Ukrainian defense industry difficult. While a joint venture with or investment in a particular firm may be attractive even without understanding the structure of UkrOboronProm, the perception that the consortium simply represents another layer of bureaucracy deters investment. There also needs to be more transparency vis-à-vis the assets, financial data, and technology of individual businesses.
- **Satisfying domestic demand.** There is a perception among Ukrainian military and foreign officials that UkrOboronProm is not meeting the demands of the Ukrainian military. This may be due in part to disruptions from the severing of ties between the Russian and Ukrainian defense industries. Nevertheless, given the importance of the domestic market for any defense company, potential partners will need to be able to evaluate UkrOboronProm's ability to meet Ukraine's needs as one metric of the potential value of collaboration.

Procurement and Requirements Process

As referenced in Chapter Four, there exists a widespread view that the requirements and procurement process is flawed. These flaws limit the possibility for investment and joint ventures with foreign companies by reducing the transparency of the Ukrainian procurement process. Given the current system, there are also significant questions about whether foreign producers can realistically compete for the Ukrainian market.

Strategic Trade Controls

Ukraine has a fairly well-developed strategic trade control system that includes systems for licensing and controlling the export of defense and dual-use items. However, there are significant gaps in Ukraine's regulation and enforcement of trade involving defense and dual-use items that could prevent regulators in potential partner countries, especially the United States, from approving transfer of defense items or technologies.

- The government of Ukraine lacks the ability to regulate the transfer of intangible dual-use or defense technologies, including skills, knowledge, documentation, or other forms of nonpublic information related to defense technologies. Internal compliance programs, which might better protect intellectual property and control of intangible technology transfers, are underdeveloped. Without trust in Ukraine's ability to protect intellectual property, Western defense industry partners will likely be unable to transfer proprietary technologies to operations in Ukraine.

- There are gaps in Ukraine's strategic control legislation and institutions. These include the absence of requirements for permits for imports or transit of many dual-use items; the absence of "catch-all controls," meaning the ability to control items that do not appear on control lists but are nevertheless used for military purposes; and gaps in the regulation of the trans-shipment of defense or dual-use items through Ukraine.
- Ukraine is currently not in control of all of its borders. While part of Ukraine's challenge relates to the ATO in Donetsk and Luhansk, other parts of the border are also insecure. The current separation of the State Border Guard Service from the customs service and the focus of customs on revenue collection rather than law enforcement degrade Ukraine's ability to control its borders.

Conclusion

Ukraine's processes, procedures, and institutions impose significant impediments for increased defense-technical cooperation. While problems identified with internal coordination, end-use monitoring, and flawed systems for military imports do not necessarily prevent countries from offering assistance or Ukraine from purchasing defense items, correcting these would likely facilitate greater cooperation, especially for high-tech items. The impediments to joint ventures with foreign companies are more substantial, and, as the next section explains, will take a concerted effort to address.

Recommendations to Address the Challenges to Defense-Technical Cooperation

A number of changes can help facilitate greater defense-technical cooperation with Ukraine. There is not necessarily a "one for one" solution for each of the problems listed—some solutions will help several problems. We present several suggested reforms, some that can be undertaken immediately and others that will require more time. However, rapid action by the Ukrainian government on some of the reforms would send a powerful signal, which in itself could further encourage outside assistance.

On their own, these solutions are insufficient to enable greater defense-technical cooperation—substantial improvements in the overall business climate are also necessary. Hence, the economic reforms that Ukraine is introducing, including those that are part of the EU-Ukraine Association Agreement, are also important for facilitating greater defense-technical cooperation. Reforms in areas such as taxation, deregulation, the judiciary, and public procurement can help address partial enforcement of Ukrainian regulations, reduce risk for foreign investors in joint ventures, and otherwise boost confidence in the Ukrainian government's capacity to implement agreements.

Set Up "Board to Coordinate Foreign Defense Assistance"

Lack of internal coordination on assistance is a major irritant for countries seeking to aid Ukraine. We recommend the creation of a long-term planning and capability development function in the MoD, which is the eventual solution to this problem. To address this problem in the short term, before overall defense reform can be implemented, an ad hoc "Board to Coordinate Foreign Defense Assistance" should be established to review and coordinate requests. This committee should be part of the organizational structure of the NSDC. During periods of intense international collaboration, it should meet no less than every two weeks and should include all relevant organizations from the MoD and GS (International Cooperation, Armaments, Logistics, and, possibly, recent field commanders who can provide a "reality check" on requests) and from the other security organizations, including the Ministry of Interior, National Guard, the SBU, and the State Border Guard Service. These meetings should also include representatives from Ukrainian customs and the Prime Minister's office. The board should meet regularly with relevant donors and coordinate closely with the Multilateral Joint Commission for Defense Reform and Security

Cooperation with Ukraine, which works to help prioritize assistance and training needs and identify appropriate providers.

The committee should have its own staff, run by a midlevel official working directly for the Secretary of the NSDC. The staff of the committee should maintain records of all meetings, all submissions of requests for assistance to donor nations, and a rolling schedule of imports of defense items. The staff should also establish the agenda for each meeting, ensure appropriate attendance, and act as a clearinghouse of information and point of contact for donor nations to consult in the event of questions or concerns.

The charter of the committee should be drafted as soon as possible and approved by the NSDC. This charter should have a specific expiration date (we recommend expiration after two years) to ensure that it is ad hoc and not a permanent committee, with an option for a two-year renewal if circumstances demand it.

Establishing this committee quickly would signal to donors that Ukraine takes their concerns about lack of coordination seriously. It would, in part, allay concerns that Ukraine does not adequately appreciate foreign assistance. It also would help establish habits of interagency cooperation that Ukraine should institutionalize and would pave the way for the establishment of ad hoc committees on a variety of other topics requiring much closer coordination.

Undertake Suggested MoD Reforms

Building on recommendations for reform of the MoD and GS, we recommend that the Department for Capability Development (which we recommend creating earlier in this report), in collaboration with other offices and units, define the necessary capabilities that the Armed Forces of Ukraine require. It should then work with the GS and the armed services to determine whether the items could be procured domestically or would need to be acquired from abroad. The consolidation of two separate departments within the MoD and GS for international cooperation under the Strategy and Policy Department and J-5 Strategy, Planning, and International Cooperation, respectively, would reduce duplication and address the lack of coordination in requests made to foreign donors.

Following these changes, a new set of specified working committees, meeting at regular intervals, could be established to streamline and coordinate the process of requesting assistance from foreign countries. Donor countries would also find it easier to deal with offices and functions that are similar to their own in the reformed Ukrainian MoD, especially if it resulted in increased transparency with donor nations.

Set Up a Pilot Project for End-Use Monitoring and Accountability

Although the Armed Forces of Ukraine do have a system for tracking equipment, it is largely paper-based. The information is neither available over an electronic network nor readily accessible by donor nations.

In coordination with existing pilot projects led by NATO, Ukraine should consider a pilot project to electronically track donated items, with a focus on items that are expensive, technically sensitive, and require additional safeguards according to licensing agreements (e.g., night-vision devices). While an electronic warehouse management system that is fully integrated into the logistics and acquisition system is needed in Ukraine, a short-term pilot project might demonstrate the advantages of an electronic tracking system and pave the way for longer-term reforms.

Such an effort would result in better tracking of sensitive defense items provided by donor nations. It would signal to those nations Ukraine's seriousness in establishing accountability and in tracking transferred items. Without both a functioning system and a sense that Ukraine takes the task seriously, many Western nations may be unwilling to risk providing Ukraine with anything more sophisticated and sensitive than they already have.

Facilitate Defense Imports Outside of UkrOboronProm

We recommend eliminating UkrOboronProm's exclusive control of imported defense items and giving the MoD explicit, streamlined authority to conduct foreign procurement in an expedited process for all items that are immediately needed (including armaments), within a specified budget. While the MoD has already adopted e-procurement processes for supplies required for the current conflict, these processes do not apply to armaments, and regulatory requirements continue to make it impractical for foreign companies to bid on military tenders. By adopting a streamlined process that applies to armaments, the Ukrainian government would be able to buy the critical military equipment and weapons it needs from abroad, such as secure radios and equipment for improved reconnaissance, without delays or "surcharges" placed on it by UkrOboronProm. Facilitating foreign procurement would also force the defense firms subordinate to UkrOboronProm to be more competitive, more efficient, and better at meeting Ukraine's military needs.

Giving increased authority to the MoD for immediate procurement needs will necessarily be a stopgap measure. A comprehensive solution for introducing foreign competition to tenders will require creating a single procurement department, which would have the authority to purchase from both foreign and Ukrainian suppliers. While some may argue that procurement of defense items should be left to organizations with close ties to and knowledge of the defense industry, such organizations are far more likely to have conflicts of interest that prevent the efficient, affordable, and rapid fulfillment of Ukraine's defense needs.

Ease Rules on Imports of Donated Defense Items as Part of a Comprehensive Reform of Customs and Border Protection

Representatives from Ukraine's customs authority need to be included in interagency discussions about requests for Western assistance. Customs representatives should be part of the Board to Coordinate Foreign Defense Assistance, and need to be notified of deliveries and instructed to expedite them to their intended recipients. Changing the laws surrounding the import of foreign military assistance will also be important to address questions about the need to pay customs duties on foreign military assistance. One alternative is to give the President or Prime Minister greater authority to waive customs duties; such authority currently lies with the Verkhovna Rada.

In the longer term, Ukraine's ability to import high-tech equipment and develop cooperation with Western defense companies will be limited by its inability to demonstrate control over its borders. In addition to working to secure its borders, Ukraine should consider merging the customs administration with the Border Police.

Continue to Improve Strategic Trade Controls

While Ukraine has made substantial progress in developing the State Service for Export Control and has a strong export-licensing system, additional measures are needed to improve Ukraine's strategic trade controls in order to persuade Western partners of the feasibility of greater defense-related economic partnerships. Close cooperation with Western partners seeking to help Ukraine develop its strategic trade controls, including through the U.S. Export Control and Related Border Security Program and efforts led by the EU and its member states, will be critical to building Ukraine's reputation as a safe destination for technology transfer. To this end, the Ukrainian government should

- increase its legal authority to regulate the transfer of intangible dual-use and defense technologies, including adding support for developing internal compliance programs in Ukraine's defense industry
- develop "catch-all" controls and increase regulation over the trans-shipment of dual-use and defense items.

Reform UkrOboronProm

Discussions with Ukrainian and Western officials frequently included criticisms of UkrOboronProm, which appears to have been set up and organized to facilitate closer defense cooperation with Russia and the Russian defense industry, in addition to its goal of consolidating defense-related state-owned enterprises owned by different government entities. As close cooperation with the Russian defense industry has ceased, some of the features of UkrOboronProm that facilitated cooperation with Russia have now become liabilities. Its structure and operations will need to be made more transparent if Ukraine hopes to develop cooperation between its defense establishment and Western defense firms.

We recommend that the Ukrainian government take the following steps to improve the transparency, efficiency, and competitiveness of its defense industrial sector, which will also make defense industries in Ukraine more-attractive partners for foreign defense companies:

1. UkrOboronProm's holdings in its subsidiaries and affiliates need to be made a matter of public record. It needs to publish

an annual report with consolidated accounts so that revenues, costs, profits, and losses can be clearly tracked. It also needs to publish annual audit reports.

2. Unconsolidated subsidiaries and all auxiliary companies need to be fully incorporated as public companies. They need to have boards of directors, publish annual reports, and conduct and publish annual audit reports. Major shareholders need to be fully disclosed, including their ownership stakes.
3. After careful review, the Ukrainian government needs to decide whether to privatize, liquidate, or retain state-controlled enterprises in the defense industry. As part of this review, the Ukrainian government will need to decide whether a consolidated company like UkrOboronProm could usefully manage enterprises that remain under state control or whether these companies would be more efficient if they ran their affairs independently. Depending on the outcome of this review, streamlining regulation of joint ventures involving state-owned enterprises might be appropriate. Adopting standard business practices will also be an important part of these reforms.

Conclusion

Ukraine has the ability to remove or reduce the various impediments to greater defense-technical cooperation. Implementing quick-impact measures, such as establishing the Board to Coordinate International Assistance, is critical to building momentum for reform and gaining greater trust among Ukraine's partners. A record of improved cooperation in one area, such as facilitating greater international assistance, would build trust that would carry over to other areas, such as joint ventures. Many of the problems that Ukraine faces with defense-technical cooperation with the West stem from legacy institutions that were designed to facilitate close cooperation with Russia or from perceptions of corruption. Confronting vested interests; bringing greater transparency to Ukraine's defense establishment; and systematically reconsidering the structure of UkrOboronProm, the MoD, and other institutions will be highly beneficial to enabling greater cooperation with Western partners.

None of these reforms will guarantee greater cooperation. Many countries are concerned about the political and strategic risks of supporting Ukraine. Even if Ukraine takes steps to improve safeguards, potential source countries may still be concerned about the risks of retransfer and misuse. The global market for defense items is highly competitive, and foreign defense firms are likely to be skeptical of supporting potential competitors within a small market for defense exports. Nevertheless, the above reform measures offer Ukraine the best-available means to build trust with its foreign partners and encourage greater defense-technical cooperation in the future.

CHAPTER SEVEN

Conclusions

The Ukrainian security institutions that existed in March 2014 were unable to respond effectively to the emerging conflict in Eastern Ukraine. In many cases, these institutions were not designed to be effective at warfighting or ensuring internal security. They were inefficient at using resources; in some cases, highly corrupt; and did not meet Euro-Atlantic standards for democracy, the rule of law, or civilian control of the military.

Since March 2014, the Ukrainian security establishment has made significant progress. A mobilization system has met immediate needs for military personnel; the logistics system has improved in its ability to provide supplies to the troops; a major effort is under way to reform the MIA; and there have been improvements in C3I. These efforts are the result of hard work by the Armed Forces of Ukraine, Ukraine's police and leadership, other civilian staff, and volunteers.

Nevertheless, these efforts have been insufficient to address the current and future threats facing Ukraine. The solutions put in place tend to be superficial or ad hoc. The fundamentals of the pre-Maidan system generally remain in place. The overall system needs substantial reform to enable Ukraine's security sector to be effective, efficient, transparent, and accountable.

This report defines a road map for security sector reform by providing a range of recommendations for organizational change in line with Euro-Atlantic standards and approaches. Our most important recommendations for reform are these:

- Define responsibilities and authorities of the security sector leadership. The President should have responsibility for the defense of Ukraine against threats to its sovereignty and independence. Presidential responsibilities extend to the command and control of military operations and to policy control over the MoD and GS. The Minister of Defense should be the senior official charged with making and carrying out the Ukrainian government's policy on defense; the chain of command should run from the President to the Minister of Defense, to the CHoD, to the JOC.
- Improve coordination across the government by expanding the responsibilities of the NSDC to include implementation of the President's decisions, an expansion of the role of the JCOI, the creation of a new cybersecurity and defense structure under a JCOCS, and the setting up of a committee to coordinate foreign defense assistance.
- Define the SBU as a domestic intelligence organization, with authorities that are defined more clearly and more narrowly, while retaining responsibility for some law enforcement activities in coordination with other agencies.

While there is no perfect system for Ukraine, we believe our recommendations offer a significant improvement over what currently exists, and will be robust across a range of possible contingencies. Most importantly, Ukraine needs to ensure that its security organizations are accountable to the public, integrated into society, and economically sustainable.

We offer recommendations that can be implemented immediately and that should have immediate impact, along with larger-scale reform efforts that will require further study and more time. Both are critical. Immediate action is important to build momentum, show that reform is possible, and fix problems that are undermining success. But large-scale, complex organizational change is also necessary for Ukraine to achieve its goals, although it will require significant discussion, consensus-building, and political support.

Implementation of our recommendations will be challenging. There are many factors favoring inertia: A complex set of existing laws and regulations; government officials comfortable with current practices; questions about the necessity of reform; and, for some, hesitation about Ukraine's increasing integration into Euro-Atlantic organizations. Ukraine is also fighting a war while it seeks to reform, so organizations must change while still fulfilling core functions.

Despite these challenges, reform is possible. Many countries, such as Poland and the Czech Republic, have significantly changed their Soviet-styled institutions to conform to NATO standards. Many have undergone significant reform in the midst of conflict, as the United States did during World War II. Indeed, a period of conflict may be a good time for reform, as it clearly demonstrates the imperative to change, and can offer a sharp break from past practices.

Ultimately, implementing reform of the security sector is in the hands of the leaders of Ukraine. Achieving reform will require support from the governing coalition of Ukraine, from the existing bureaucracy to ensure that institutions continue to function, and civil society. The international community can provide continued assistance, both through top-down encouragement of reform, as well as bottom-up support to help change bureaucratic cultures. International partners can also offer advice in technical areas, such as project management, personnel, and finance. However, key reforms can only be implemented by the Ukrainian government: changing laws, reorganizing departments, recruiting new personnel, and developing new training programs. Ukrainian senior leadership is essential to ensure that reforms fit Ukraine's culture, and avoid repeating past mistakes. Ukraine's government and Ukrainian society have an opportunity to develop institutions that will break with the past, secure the country's future, and plot a new trajectory toward Euro-Atlantic integration. We hope that the recommendations contained here will help with this process.

Acknowledgments

We would like to thank Dmytro Shymkiv for initiating this study and for supporting it through to its completion. Oleksandr Lytvynenko oversaw our work for the National Security and Defense Council of Ukraine and always provided wise counsel and insights. Archil Tsintsadze has been our chief point of contact in Ukraine through most of this project. Without his support and assistance in setting up meetings, providing feedback, and deciphering Ukraine's current system and needs, this report could not have been completed. Dmytri Los was a great help in the project's early phases. Vyacheslav Hnatyuk served as our interpreter in Ukraine and translated draft chapters and briefings into Ukrainian. His work was extraordinary. We greatly appreciate their help.

Within the RAND Corporation, Olesya Tkacheva contributed to our understanding of the issues of transparency and accountability in Ukraine. William Young, Andrew Liepman, and Sina Beaghley wrote an excellent report on intelligence reform from which we drew heavily for what we present here. We thank them for their insights. David Gompert carefully reviewed the chapters on intelligence and cybersecurity and provided excellent guidance and comments. Bernard Rostker provided crucial advice and guidance on military and civilian personnel policy. Svitlana Kobzar repeatedly traveled to Ukraine, offered analytic support, wrote sections of several reports, and provided invaluable insight into Ukraine's political system. Katya Migacheva conducted fieldwork in Ukraine, including research on the current Ukrainian security sector and wrote sections of the overall national security sector report. Jan Gaspers, Nathan Ryan, Caolionn O'Connell, and Isaac Porche provided invaluable research support and analysis in the drafting of the cyber chapter. Our thanks also go to Olena Bogdan and Etienne Rosas for their excellent research support.

We would also like to thank officials and officers at the European Commission, EUAM Ukraine, North Atlantic Treaty Organization (NATO) headquarters and the NATO Liaison office in Kyiv, the U.S. State Department, and the U.S. Department of Defense (including the Department of Defense–funded Defense Institution Building team) for the insights provided by staff engaged in assisting Ukraine with security sector reform. Phil Jones, the United Kingdom defense adviser in Ukraine, offered critical advice and insight, as did Glen Grant of the Defense Institution Building Team. U.S. Ambassador to Ukraine Geoffrey R. Pyatt, Colonel Cynthia Matuskevich, and Douglas Hoyt of the U.S. Embassy in Kyiv have been especially supportive of the project. Of course, any errors are our own.

Our report benefited from thoughtful and helpful reviews from Ambassador John Herbst and Stephen Flanagan. We would also like to thank Blair Smith and Arwen Bicknell for their editorial support in the report's publication.

We would particularly like to thank the many Ukrainians engaged in the security sector who have given so freely of their time and insights. Ultimately, it is their efforts that will determine the success of any recommendations adopted from this report.

Abbreviations

ATO	Anti-Terror Operation
C3I	command, control, communications, and intelligence
CERT	Computer Emergency Response Team
CERT-UA	Computer Emergency Response Team-Ukraine
CHoD	Chief of Defense Force
CoM	Cabinet of Ministers
EU	European Union
GS	General Staff
ICT	information communication technology
JCOCS	Joint Committee on Cyber Security
JCOI	Joint Committee on Intelligence
JOC	Joint Operational Command
MIA	Ministry of Internal Affairs
MoD	Ministry of Defense
NATO	North Atlantic Treaty Organization
NSDC	National Security and Defense Council of Ukraine
SBU	Security Service of Ukraine
SSSCIP	State Services for Special Communication and Information Protection

Bibliography

Arms Control Association, "The Missile Technology Control Regime at a Glance," December 2012. As of September 11, 2015:
https://www.armscontrol.org/factsheets/mtcr

Australian Department of Defence, "Exercise: Talisman Sabre: Home," 2005. As of February 28, 2016:
http://web.archive.org/web/20050713122123/http://www.defence.gov.au/talisman_sabre/

Aziz, Sahar, *U.S. Foreign Aid and Morsi's Ouster*, Middle East Institute, July 31, 2013. As of July 22, 2015:
http://www.mei.edu/content/us-foreign-aid-and-morsis-ouster

Azulai, Yuval, "Interview with the Man Responsible for Israeli Arms Exports," [in Hebrew] *Globes*, June 7, 2015. As of September 11, 2015:
http://www.globes.co.il/news/article.aspx?did=1001042515

Ball, Sam, "Arms Sales Becoming France's New El Dorado, But at What Cost?" *France 24*, May 4, 2015. As of August 5, 2015:
http://www.france24.com/en/20150503-arms-sales-becoming-france-new-el-dorado-but-what-cost-francois-hollande-saudi-arabia-rafale

Banham, Mark, "UK Government to Invest a Further £2bn to Police Cyberspace," *International Business Times*, 2015. As of August 5, 2015:
http://www.ibtimes.co.uk/government-invest-further-2bn-police-cyberspace-1515691

Benn, Aluf, "Israel Selling Military Wares to Mideast Countries, Britain Says," *Haaretz*, June 11, 2013. As of September 11, 2015:
http://www.haaretz.com/news/diplomacy-defense/.premium-1.528993

Cabinet Office, *The UK Cyber Security Strategy: Report on Progress and Forward Plans*, December 2014. As of February 28, 2016:
https://www.gov.uk/government/uploads/system/uploads/attachment_data/file/386093/The_UK_Cyber_Security_Strategy_Report_on_Progress_and_Forward_Plans_-_De___.pdf

Cavaiola, Lawrence J., David C. Gompert, and Martin Libicki, "Cyber House Rules: On War, Retaliation and Escalation," *Survival*, Vol. 57, No. 1, 2015.

Cohen, Gili, "Israeli Arms Exports Down $1 Billion in 2014," *Haaretz*, May 21, 2015. As of September 11, 2015:
http://www.haaretz.com/news/diplomacy-defense/.premium-1.657613

Constitution of Ukraine, Kyiv: Supreme Council of Ukraine, 1996.

Council of the European Union, "European Union Code of Conduct on Arms Exports," 8675/2/98 Rev. 2, June 5, 1998.

———, "Council Common Position 2003/468/CFSP," *Official Journal of the European Union*, June 23, 2003. As of October 1, 2015:
http://eur-lex.europa.eu/legal-content/EN/TXT/HTML/?uri=CELEX:32003E0468&from=EN

———, "Council Common Position 2008/944/CFSP," *Official Journal of the European Union*, L335, December 8, 2008.

———, "Regulations," *Official Journal of the European Union*, No. 428/2009, May 5, 2009a. As of October 1, 2015:
http://eur-lex.europa.eu/LexUriServ/LexUriServ.do?uri=OJ:L:2009:134:0001:0269:en:PDF

———, "Directive 2009/43/EC of the European Parliament and of the Council," *Official Journal of the European Union*, May 6, 2009b. As of August 26, 2015:
http://eur-lex.europa.eu/LexUriServ/LexUriServ.do?uri=OJ:L:2009:146:0001:0036:en:PDF

———, "Notices from European Union Institutions, Bodies, Offices and Agencies," *Official Journal of the European Union*, Vol. 57, April 9, 2014. As of October 1, 2015:
http://eur-lex.europa.eu/legal-content/EN/TXT/PDF/?uri=OJ:JOC_2014_107_R_0001&from=EN

Defence Committee, Parliament of the UK, "Defence and Cyber-Security—Defence Committee Contents," January 9, 2013. As of February 28, 2016:
http://www.publications.parliament.uk/pa/cm201213/cmselect/cmdfence/106/10605.htm

Defense Export Controls Agency, "About DECA," Government of Israel, undated-a. As of July 22, 2015:
http://www.exportctrl.mod.gov.il/ExportCtrl/ENGLISH/About+DECA/

———, "Defense Export Control Policy," Government of Israel, undated-b. As of July 22, 2015:
http://www.exportctrl.mod.gov.il/ExportCtrl/ENGLISH/Defense+Export+Control+Policy/

Defense Institute of Security Assistance Management, *The Management of Security Cooperation (Green Book)*, 2015.

Defense Security Cooperation Agency, "Excess Defense Articles (EDA)," undated. As of July 22, 2015:
http://www.dsca.mil/programs/excess-defense-articles-eda

———, *Security Assistance Management Manual*, 2015. As of July 22, 2015:
http://www.samm.dsca.mil/

Demper, Alexandra, "UK Implements Changes to EU Common Military List; UK Military List Replaced," Sanctions Update, March 27, 2015.

Directorate of Defense Trade Controls, "Compliance Program Guidelines," U.S. Department of State, 2015a. As of July 22, 2015:
https://www.pmddtc.state.gov/compliance/documents/compliance_programs.pdf

———, "The International Traffic in Arms Regulations: ITAR, 22 CFR 120-130," U.S. Department of State, updated July 7, 2015b. As of July 22, 2015:
https://www.pmddtc.state.gov/regulations_laws/itar.html

Dyson, E., G. Gilder, G. Keyworth, and A. Toffler, "Cyberspace and the American Dream: A Magna Carta for the Knowledge Age," *Information Society*, Vol. 12, No. 3, 1996. As of February 25, 2016:
http://www.tandfonline.com/doi/pdf/10.1080/019722496129486.

"Electronic Warfare," TheyWorkForYou.com, 2014. As of November 26, 2015:
http://www.theyworkforyou.com/wrans/?id=2014-09-04a.207634.h

"EU Defense Procurement and Export Control Policies," web page, July 14, 2014. As of October 1, 2015:
http://www.export.gov/europeanunion/defenseprocurement/index.asp

European Network Against Arms Trade, *European Export Credit Agencies and the Financing of Arms Trade*, 2007.

European Parliament, Council of the EU, European Economic and Social Committee, and Committee of the Regions, *Cybersecurity Strategy of the European Union: An Open, Safe and Secure Cyberspace*, Joint Communication to the European Parliament, the Council, the European Economic and Social Committee and the Committee of the Regions, 2013. As of September 20, 2015:
http://eur-lex.europa.eu/legal-content/EN/TXT/?uri=CELEX:52013JC0001

European Union Agency for Network and Information Security, *A Step-By-Step Approach on How to Set Up a CSIRT Including Examples and a Checklist in Form of a Project Plan*, Heraklyon, 2006.

———, *An Evaluation Framework for National Cyber Security Strategies*, Heraklyon, 2014.

Farmer, Ben, "Fitness Tests Waived for MoD's New Reservist Cyber Warriors," *The Telegraph*, 2015. As of November 26, 2015:
http://www.telegraph.co.uk/news/uknews/defence/11360976/Fitness-tests-waived-for-MoDs-new-reservist-cyber-warriors.html

Federal Bureau of Investigation, "FBI—Cyber Task Forces," 2015.

Fergusson, Ian F., and Paul K. Kerr, *The U.S. Export Control System and the President's Reform Initiative*, Congressional Research Service, January 13, 2014. As of July 29, 2015:
https://www.fas.org/sgp/crs/natsec/R41916.pdf

Fluri, Philipp, Marcin Koziel, and Andrii Yermolaiev, eds., *The Security Sector Legislation of Ukraine*, Kyiv, Ukraine: Center for Army, Conversion and Disarmament Studies, 2014.

Fluri, Philipp, and V. G. Radetskiy, *Security Sector Reform in Ukraine: Quo Vadis?* Kyiv: National Defense Academy of Ukraine and Geneva: Centre for the Democratic Control of Armed Forces, 2010. As of January 29, 2016:
http://www.dcaf.ch/Publications/Security-Sector-Reform-in-Ukraine-Quo-Vadis

Global Cyber Security Capacity Centre, "Cyber Security Capability Maturity Model (CMM)—V1.2," University of Oxford, 2014. As of February 25, 2016:
https://www.sbs.ox.ac.uk/cybersecurity-capacity/system/files/CMM%20Version%201_2_0.pdf

Gould, Joe, "Constructing a Cyber Superpower," *DefenseNews*, 2015. As of December 28, 2015:
http://www.defensenews.com/story/defense/policy-budget/cyber/2015/06/27/us-cyber-command-budget-expand-fort-meade-offensive/28829321/

Government of Israel, "Import and Export Order (Control of Dual-Purpose Goods, Services and Technology Exports)," 5766–2006, August 23, 2006.

Green, James A., *Cyber Warfare: A Multidisciplinary Analysis*, New York: Taylor & Francis, 2015.

Gross, Judah Ari, "As South Sudan Bloodies Itself, Israeli Arms Sales Questioned," *Times of Israel*, August 20, 2015. As of September 11, 2015:
http://www.timesofisrael.com/as-south-sudan-bloodies-itself-israeli-arms-sales-questioned/

Hansen, Susanne Therese, and Nicholas Marsh, "Normative Power and Organized Hypocrisy: European Union Member States' Arms Export to Libya," *European Security*, Vol. 24, No. 2, 2015.

Headington, Yvonne, "MoD's Cyber Security Capabilities," 2013. As of November 30, 2015:
http://www.battle-technology.com/exhibitions.asp?key=604

Her Majesty's Treasury, Government Communications Headquarters, and George Osborne, "Chancellor's Speech to GCHQ on Cyber Security," 2015. As of November 26, 2015:
https://www.gov.uk/government/speeches/chancellors-speech-to-gchq-on-cyber-security

Holm, Kyrre, "Europeanising Export Controls: The Impact of the European Union Code of Conduct on Arms Exports in Belgium, Germany, and Italy," *European Security*, Vol. 15, No. 2, 2006, pp. 213–234.

"Israeli Defense Industry Exports Under Scrutiny," United Press International, July 19, 2013. As of July 22, 2015:
http://www.upi.com/Business_News/Security-Industry/2013/07/19/Israeli-defense-industry-exports-under-scrutiny/UPI-11581374259134/

"Joint Cyber Reserve," TheyWorkForYou, 2015. As of November 26, 2015:
http://www.theyworkforyou.com/wrans/?id=2015-02-25.225462.h

Joint Forces Command, "Working for JFC," 2015. As of November 26, 2015:
https://www.gov.uk/government/organisations/joint-forces-command/about/recruitment

Klimburg, Alexander, ed., *National Cyber Security Framework Manual*, Tallin: NATO Cooperative Cyber Defence Centre, December 2012.

Knight, Ben, "Germany's Arms Exports Turn into Gamesmanship," *Deutsche Welle*, April 16, 2014. As of August 5, 2015:
http://www.dw.com/en/germanys-arms-exports-turn-into-gamesmanship/a-17566710

Kramer, Andrew, and David Herszenhorn, "Ukrainian Soldiers' Retreat from Eastern Town Raises Doubt for Truce," *New York Times*, February 18, 2015. As of August 8, 2016:
http://www.nytimes.com/2015/02/19/world/europe/ukraine-conflict-debaltseve.html

Law of Ukraine, "On the Security Service of Ukraine," Article 1, 1992.

———, "On Democratic Civilian Control of State Military Organisation and Law Enforcement Bodies," Article 13.2, 2003.

———, "On Democratic Civilian Control of State Military Organisation and Law Enforcement Bodies," Article 15, 2003.

———, "On Counterintelligence Activity," Article 1, 2005.

Law of the United Kingdom, "Security Service Act of 1989," Chapter 5, Section 1, 1989. As of February 25, 2016:
http://www.legislation.gov.uk/ukpga/1989/5/section/1

Leigh, David, and Rob Evans, "BAE Admits Guilt Over Corrupt Arms Deals," *The Guardian*, February 5, 2010. As of August 7, 2015:
http://www.theguardian.com/world/2010/feb/05/bae-systems-arms-deal-corruption

Levitsky, Steven, and Lucan Way, *Competitive Authoritarianism: Hybrid Regimes After the Cold War*, New York: Cambridge University Press, 2010.

Libicki, Martin C., David Senty, and Julia Pollak, *Hackers Wanted: An Examination of the Cybersecurity Labor Market*, Santa Monica, Calif.: RAND Corporation, RR-430, 2014. As of December 28, 2015:
http://www.rand.org/pubs/research_reports/RR430.html

Lutterbeck, Derek, *The Paradox of Gendarmeries: Between Expansion, Demilitarization and Dissolution*, The Geneva Centre for the Democratic Control of Armed Forces, 2013. As of August 8, 2016:
http://www.dcaf.ch/content/download/150865/2344885/file/SSR_8_EN.pdf

Manor, Hadas, "Israel and US to Review Arms Export Controls," *Globes*, March 8, 2006. As of September 11, 2015:
http://www.globes.co.il/en/article-1000069319

Marks, Paul, "UK Will Launch Its Own Cyberattacks, Not Just Defend," *New Scientist*, October 1, 2013. As of September 11, 2015:
https://www.newscientist.com/article/dn24304-uk-will-launch-its-own-cyberattacks-not-just-defend/

Minakov, Mikhail, "A Decisive Turn? Risks for Ukrainian Democracy After the Euromaidan," Carnegie Endowment for International Peace, Washington, D.C., February 3, 2016. As of July 28, 2016:
http://carnegieendowment.org/2016/02/03/decisive-turn-risks-for-ukrainian-democracy-after-euromaidan/itf4

Ministry of Defence and Joint Forces Command, "Joint Forces Command Reaches Full Operating Capability," April 2, 2013. As of February 28, 2016:
https://www.gov.uk/government/news/joint-forces-command-reaches-full-operating-capability

Ministry of Defence, Joint Forces Command, and The Rt Hon Philip Hammond MP, "New Cyber Reserve Unit Created," September 29, 2013. As of February 28, 2016:
https://www.gov.uk/government/news/reserves-head-up-new-cyber-unit

National Information Assurance Partnership, home page, undated. As of December 23, 2015:
https://www.niap-ccevs.org/

National Security Agency, "The NSA/CSS Mission—NSA/CSS," April 15, 2011. As of November 26, 2015:
https://www.nsa.gov/about/mission-strategy/

NATO—*See* North Atlantic Treaty Organization.

Norton-Taylor, Richard, "Britain Plans Cyber Strike Force—With Help From GCHQ," *The Guardian*, 2013. As of November 30, 2015:
http://www.theguardian.com/uk-news/defence-and-security-blog/2013/sep/30/cyber-gchq-defence

North Atlantic Treaty Organization, "NATO's Practical Support to Ukraine," Fact Sheet, June 2015. As of July 30, 2016:
http://www.nato.int/nato_static_fl2014/assets/pdf/pdf_2015_06/20150624_1506-Factsheet_PracticalSupportUkraine_en.pdf

Orttung, Robert W., "What Hinders Reform in Ukraine?" George Washington University Elliott School of International Affairs, Washington, D.C., PONARS Eurasia Policy Memo No. 166, September 2011. As of July 28, 2016:
http://www.ponarseurasia.org/sites/default/files/policy-memos-pdf/pepm_166.pdf

Parkin, Brian, "German Split Opens Over Blocked Sale of Tanks to Saudi Arabia," *Bloomberg Business*, March 8, 2015. As of August 5, 2015:
http://www.bloomberg.com/news/articles/2015-03-08/german-split-opens-over-blocked-sale-of-tanks-to-saudi-arabia

"Parliament Simplified Import of Medicines and Bulletproof Jackets to Ukraine" [in Ukrainian], *Ukr.Media*, July 1, 2014. As of August 5, 2015:
https://ukr.media/ukrain/206484/

Petrov, Oleksii, "Political and Budgetary Oversight of the Ukrainian Intelligence Community," thesis, Monterey, Calif.: Naval Postgraduate School, September 2007.

Phalnikar, Sonia, "Europe Tries to Reconcile Libya Criticism with Booming Arms Exports," *Deutsche Welle*, February 24, 2011. As of August 5, 2015:
http://www.dw.com/en/europe-tries-to-reconcile-libya-criticism-with-booming-arms-exports/a-14872650

Pomper, Miles A., "U.S., Israel Reach China Arms Deal," *Arms Control Today*, September 1, 2005. As of September 11, 2015:
https://www.armscontrol.org/act/2005_09/USIsraelChinaDeal

Rapaport, Amir, "New US-Israel Crisis Involving Defense Exports to China," *Israel Defense*, December 21, 2013. As of September 11, 2015:
http://www.israeldefense.co.il/en/content/new-us-israel-crisis-involving-defense-exports-china

Reznikova, O., and V. Tsiukalo, *Development of Ukrainian National Security Strategic Planning and Forecasting System*, Kyiv, Ukraine: National Institute for Strategic Studies, June 2015. As of December 30, 2015:
http://en.niss.gov.ua/articles/529/

Robinson, Neil, Luke Gribbon, Veronika Horvath, and Kate Robertson, *Cyber-Security Threat Characterisation: A Rapid Comparative Analysis*, Santa Monica, Calif.: RAND Corporation, RR-235-CATS, 2013. As of December 28, 2015:
http://www.rand.org/pubs/research_reports/RR235.html

Saferworld and China Arms Controls and Disarmament Association (CACDA), *The Evolution of EU and Chinese Arms Export Controls*, March 2012. As of September 18, 2015:
http://www.saferworld.org.uk/resources/view-resource/687-the-evolution-of-eu-and-chinese-arms-export-controls

Sawer, Patrick, "General Sir Richard Dannatt Condemns Armoured Vehicle Transfer to Ukraine," *The Telegraph*, February 14, 2015. As of September 18, 2015:
http://www.telegraph.co.uk/news/worldnews/europe/ukraine/11412838/Sir-Richard-Dannatt-condemns-armoured-vehicle-transfer-to-Ukraine.html

Schmidt, Lara, *Perspective on 2015 DoD Cyber Strategy*, Testimony presented before the House Armed Services Committee, Santa Monica, Calif.: RAND Corporation, CT-439, 2015. As of December 28, 2015:
http://www.rand.org/pubs/testimonies/CT439.html

Solivan, Douglas A., Sr., "Communications-Electronics Command Cyber Training Range Launches," United States Army: Logistics and Readiness Center, CECOM, June 23, 2015. As of December 23, 2015:
http://www.army.mil/article/150996/Communications_Electronics_Command_cyber_training_range_launches/

Stavridis, James G., and Dave Weinstein, "Divide and Conquer: Why Dual Authority at the NSA and Cyber Command Hurts U.S. Cybersecurity," *Foreign Affairs*, 2013. As of November 26, 2015:
https://www.foreignaffairs.com/articles/united-states/2013-10-22/divide-and-conquer

Sternstein, Aliya, "The Military's Cybersecurity Budget in 4 Charts," *DefenseOne*, 2015a. As of November 26, 2015:
http://www.defenseone.com/management/2015/03/militarys-cybersecurity-budget-4-charts/107679/

———, "US Cyber Command Has Just Half the Staff It Needs," *DefenseOne*, 2015b. As of February 28, 2016:
http://www.defenseone.com/threats/2015/02/us-cyber-command-has-just-half-staff-it-needs/104847/

Tran, Pierre, "Egypt Seeks Bank Loans for French Arms Buy," *DefenseNews*, February 9, 2015. As of August 7, 2015:
http://www.defensenews.com/story/defense/policy-budget/industry/2015/02/06/france/22934245/

United Nations Office for Disarmament Affairs, "The Arms Trade Treaty," 2014.

U.S. Department of Defense, "Active Duty Military Personnel by Service by Rank/Grade: April 2015," DoD Personnel, Workforce Reports & Publications, Defense Manpower Data Center, 2015. As of April 15, 2015:
https://www.dmdc.osd.mil/appj/dwp/dwp_reports.jsp

———, "Military Pay Charts, Jan 1, 2015," Defense Finance and Accounting Service, 2015. As of April 15, 2015:
http://www.dfas.mil/militarymembers/payentitlements/military-pay-charts.html

U.S. Strategic Command, "U.S. Cyber Command—U.S. Strategic Command," March 2015. As of November 26, 2015:
https://www.stratcom.mil/factsheets/2/Cyber_Command/

Van der Meulen, Nicole, Eun Jo, and Stefan Soesanto, *Cybersecurity in the European Union and Beyond: Exploring the Threats and Policy Responses*, Santa Monica, Calif.: RAND Corporation, RR-1354-EP, 2015. As of November 26, 2015:
http://www.rand.org/pubs/research_reports/RR1354.html